AUTHENTIC OPPORTUNITIES FOR WRITING ABOUT MATH
in Early Childhood

Teach students to write about math so they can improve their conceptual understanding in authentic ways. This resource offers hands-on strategies you can use to help students in grades PreK–2 discuss and articulate mathematical ideas, use correct vocabulary, and compose mathematical arguments.

Part One discusses the importance of emphasizing language to make students' thinking visible and to sharpen communication skills, while attending to precision. Part Two provides a plethora of writing prompts and activities: Visual Prompts; Compare and Contrast; The Answer Is; Topical Questions; Writing About; Journal Prompts; Poetry/Prose; Cubing and Think Dots; RAFT; Question Quilts; and Always, Sometimes, Never. Each activity is accompanied by a clear overview plus a variety of examples. Part Three offers a crosswalk of writing strategies and math topics to help you plan, as well as a sample anchor task and lesson plan to demonstrate how the strategies can be integrated.

Throughout each section, you'll also find Blackline Masters that can be downloaded for classroom use. With this book's engaging, standards-based activities, you'll have young children communicating like fluent mathematicians in no time!

Tammy L. Jones has taught students from first grade through college. Currently, she is consulting with individual school districts in training mathematics teachers on effective techniques for being successful in the mathematics classroom, supporting mathematics instruction, and STEM integrations. She is co-author of two book series published with Routledge: *Strategies for Common Core Mathematics* and *Strategic Journeys for Building Logical Reasoning*.

Leslie A. Texas has over 20 years of experience working with K–12 teachers and schools across the country to enhance rigorous and relevant instruction. She believes that improving student outcomes depends on comprehensive approaches to teaching and learning. She is co-author of two book series published with Routledge: *Strategies for Common Core Mathematics* and *Strategic Journeys for Building Logical Reasoning*.

Also Available from Tammy L. Jones and Leslie A. Texas
(www.routledge.com/k-12)

Strategic Journeys for Building Logical Reasoning, K–5:
Activities Across the Content Areas

Strategic Journeys for Building Logical Reasoning, 6–8:
Activities Across the Content Areas

Strategic Journeys for Building Logical Reasoning, 9–12:
Activities Across the Content Areas

Strategies for Common Core Mathematics:
Implementing the Standards for Mathematical Practice, K–5

Strategies for Common Core Mathematics:
Implementing the Standards for Mathematical Practice, 6–8

Strategies for Common Core Mathematics:
Implementing the Standards for Mathematical Practice, 9–12

AUTHENTIC OPPORTUNITIES FOR WRITING ABOUT MATH
in Early Childhood

Prompts and Examples for Building Understanding

Tammy L. Jones and Leslie A. Texas

NEW YORK AND LONDON

Cover images: @ Getty Images

First published 2025
by Routledge
605 Third Avenue, New York, NY 10158

and by Routledge
4 Park Square, Milton Park, Abingdon, Oxon, OX14 4RN

Routledge is an imprint of the Taylor & Francis Group, an informa business

© 2025 Tammy L. Jones and Leslie A. Texas

The right of Tammy L. Jones and Leslie A. Texas to be identified as authors of this work has been asserted in accordance with sections 77 and 78 of the Copyright, Designs and Patents Act 1988.

All rights reserved. No part of this book may be reprinted or reproduced or utilised in any form or by any electronic, mechanical, or other means, now known or hereafter invented, including photocopying and recording, or in any information storage or retrieval system, without permission in writing from the publishers.

Trademark notice: Product or corporate names may be trademarks or registered trademarks, and are used only for identification and explanation without intent to infringe.

Library of Congress Cataloging-in-Publication Data
Names: Jones, Tammy L., author. | Texas, Leslie A., author.
Title: Authentic opportunities for writing about math in early childhood: prompts and examples for building understanding / Tammy L. Jones and Leslie A. Texas.
Description: New York, N.Y.: Routledge, 2025. | Includes bibliographical references.
Identifiers: LCCN 2024018921 (print) | LCCN 2024018922 (ebook) |
ISBN 9781032449289 (hardback) | ISBN 9781032445533 (paperback) |
ISBN 9781003374565 (ebook)
Subjects: LCSH: Mathematics—Study and teaching (Early childhood)—Activity programs.
Classification: LCC QA135.6 .J664 2025 (print) | LCC QA135.6 (ebook) |
DDC 808.06/6372—dc23/eng/20240630
LC record available at https://lccn.loc.gov/2024018921
LC ebook record available at https://lccn.loc.gov/2024018922

ISBN: 978-1-032-44928-9 (hbk)
ISBN: 978-1-032-44553-3 (pbk)
ISBN: 978-1-003-37456-5 (ebk)

DOI: 10.4324/9781003374565

Typeset in Warnock Pro
by codeMantra

Access the Support Material: https://resourcecentre.routledge.com/books/9781032449289 or visit https://resourcecentre.routledge.com and search for the book's ISBN, title or authors.

Special thanks to Pixaby, WordArt, and Geometer's Sketchpad for certain images used in this book.

We would like to give special thanks to Trevor Styer for his work to ensure the graphics used throughout the series were high quality and reproducible for classroom use.

Online Resources

Several of the resources in this book are available online as free downloads so you can print them for classroom use. To access them, find the book at the url below. Note that you will be asked to provide information from the book before you can obtain the downloads. https://resourcecentre.routledge.com/

You can also follow this direct link: https://resourcecentre.routledge.com/books/9781032449289

Contents

Meet the Authors — xi
Preface — xiii

PART ONE
Why Writing in Math Matters — **1**

Chapter 1   Purposeful Writing: Intentional Design — 3

PART TWO
Writing Prompts — **15**

Chapter 2   Visual Prompts — 17
Chapter 3   Compare and Contrast — 25
Chapter 4   The Answer Is… — 30
Chapter 5   Topical Questions — 38
Chapter 6   Writing About… — 47
Chapter 7   Journal Prompts — 88
Chapter 8   Poetry/Prose — 92
Chapter 9   Cubing and Think Dots — 95
Chapter 10 RAFT — 121
Chapter 11 Question Quilt — 125
Chapter 12 True/False and Always, Sometimes, and Never — 130

Contents

PART THREE
Planning and Implementation   **137**

Chapter 13   Crosswalk   139

Chapter 14   Bringing It All Together   141

Afterword   161
Bibliography   163

Meet the Authors

Collectively, Tammy and Leslie have almost 45 years of classroom experience teaching in elementary, middle, high school, and college. This has included urban, suburban, rural, and private school settings. Being active members of their professional organizations has allowed them to continually grow professionally and model lifelong learning for both their students and their peers. In their 30-plus years of combined consulting work, they have had opportunities to work with teachers and students from kindergarten through college level. This work has spanned almost all 50 states. Their work has included helping to develop standards and curriculum at the state level as well as implementing curriculum and best practice strategies at the classroom level. One of the things that sets Tammy and Leslie apart as consultants is their work with classroom teachers, modeling and offering continued support throughout the year to build capacity at the building and district levels. Tammy and Leslie co-authored the 2013 series from Eye On Education/Routledge-Taylor & Francis Group, *Strategies for Common Core Standards for Mathematics: Implementing the Standards for Mathematical Practice* (Grades K-5, 6–8, and 9–12) and the 2017 series from Routledge-Taylor & Francis Group, *Strategic Journeys for Building Logical Reasoning: Activities Across the Content Areas* (Grades K-5, 6–8, and 9–12).

An educator since 1979, **Tammy L. Jones** has worked with students from first grade through college. Currently, Tammy is consulting with individual

Meet the Authors

school districts in training teachers on strategies for making content accessible to all learners. Writing integrations as well as literacy connections are foundational in everything Tammy does. Tammy also works with teachers on effective techniques for being successful in the classroom. As a classroom teacher, Tammy's goal was that all students understand and appreciate the content they were studying; that they could read it, write it, explore it, and communicate it with confidence; and that they would be able to use the content as they need to in their lives. She believes that logical reasoning, followed by a well-reasoned presentation of results, is central to the process of learning, and that this learning happens most effectively in a cooperative, student-centered classroom. Tammy believes that learning is experiential and in her current consulting work creates and shares engaging and effective educational experiences.

Leslie A. Texas has over 25 years of experience working with K-12 teachers and schools across the country to enhance rigorous and relevant instruction. She believes that improving student outcomes depends on comprehensive approaches to teaching and learning. She taught middle and high school mathematics and science and has strong content expertise in both areas. Through her advanced degree studies, she honed her skills in content and program development and student-centered instruction. Using a combination of direct instruction, modeling, and problem-solving activities rooted in practical application, Leslie helps teachers become more effective classroom leaders and peer coaches.

Preface

A Note to Our Readers

Our previous two book series, *Strategies for Common Core Mathematics: Implementing the Standards for Mathematical Practice* (Grades K-5, 6–8, and 912), and *Strategic Journeys for Building Logical Reasoning: Activities Across the Content Areas* (Grades K-5, 6–8, 9–12), provided a set of strategies and sample tasks that teachers could implement across the curriculum to engage students at a deeper cognitive level required by the rigorous college and career ready standards.

When we took on writing this new series, we asked ourselves: What is it that teachers want and would support students in becoming better communicators of mathematics? During training with teachers on our other two series, we often were asked how teachers could get more classroom-ready materials, such as questions, writing prompts, etc., that would support their work with students on writing and reading mathematics. Therefore, we wanted to create a collection of items for educators that would be practical and versatile, easy to implement, and yield results.

For the student, we created a collection of visual prompts that provide opportunities to engage in mathematics through looking at pictures of and from the world. There is an assortment of examples supporting the academic

Preface

vocabulary associated with each math topic. Also included are ready-to-use writing prompts covering a variety of topics across the grade bands. Sets of non-typical questions are provided to promote developing a deeper understanding of mathematics. Examples of various writing styles, including creative writing, meet the needs and interests of a diverse classroom.

For educators, it is important to understand students can only become comfortable (and proficient) communicating about mathematics by practicing it regularly. Today's high-stakes assessments require students to understand mathematics in context and to explain their reasoning behind strategies and solutions. There are enough strategies included to incorporate often (daily/weekly). Using these prompts and tasks is easy once the teacher has determined the instructional goals and targeted standards for implementation. There is teacher autonomy in implementing, but the prompts and tasks are ready to be used immediately.

These are great strategies for providing a variety of ways to engage students in mathematical discourse. The materials are versatile in use as handouts, visual displays, gallery walks, electronic documents, etc. Crosswalks show examples by mathematical topic as well as by type of writing. Teachers will find strategies for authentically integrating different writing techniques in the mathematics classroom, including creative writing. A sample lesson incorporating a number of these prompts and examples is included along with unique strategies and examples for differentiation in the mathematics classroom.

PART ONE

Why Writing in Math Matters

CHAPTER 1

Purposeful Writing: Intentional Design

Communication is essential in expressing ideas clearly and effectively. Language serves as a framework for that communication. Mathematics is often said to have its' own language using symbols in addition to words. Combining mathematical language with written/spoken language can often provide deeper insight into how information is being processed, connections that are being made, conclusions drawn, etc. This data is important in assessing understanding as well as moving thinking further.

This book will look at how writing can be used in the following:

1. Making student thinking visible – formative assessment
2. Building communication skills while attending to precision – construct a viable argument and critique the reasoning of others (Standard of Mathematical Practice 3) and attend to precision (Standard of Mathematical Practice 6)
3. Establishing authentic reasons for writing, not just so we can say we did write in math.

As introduced in our book series *Strategic Journeys for Building Logical Reasoning: Activities Across the Content Areas* (Jones & Texas, 2017), there are seven opportunities for writing. These served as a guideline and informed the choices made regarding the types of writing included in this series.

Authentic Opportunities for Writing about Math in Early Childhood

- **Making Meaning** – understanding the question posed and identifying the given and needed information necessary to proceed
- **Showing Evidence** – using facts and/or data to support one's argument/hypothesis/work
- **Reflecting** – being metacognitive with respect to strategies and/or processes
- **Inquiry** – creating questions to drive investigation and/or research
- **Educating** – informing others in various forms/purposes – persuasive, descriptive, expository, and narrative
- **Creating Ideas** – brainstorming/free writing to begin framing ideas
- **Producing Products** – using products to convey a message depending on audience and purpose (research papers, proposals, brochures, essays, public service announcements, etc.)

Making Student Thinking Visible: Formative Assessment

For teachers to elicit evidence of student understanding and provide feedback that moves the learning forward, students must be able to make their thinking visible. Many students struggle to organize their thoughts and capture their thinking on paper. Starting with a straightforward tool such as the Think-Write-Pair-Share (Jones & Texas, 2017) allows student specific guidance on where to begin writing. A blank piece of paper can mean "I don't know" or "I don't care." It is an important distinction, and providing tools for students to support the "I don't know" is critical in building their capacity to help themselves. In addition, this intentional emphasis on writing highlights the importance of being a good partner by bringing something to the table when coming together to discuss ideas. Below is one example of capturing this thinking.

Think, WRITE, Pair, Share

<u>Think</u> about…

<u>WRITE</u> about what questions come to mind in the area below.

<u>PAIR</u> with your partner and discuss what each of you wrote.

Be prepared to <u>**SHARE**</u> with the whole group.

Authentic Opportunities for Writing about Math in Early Childhood

Using with a Rich Task

The following tool can be used to help students organize their thoughts around a rich task. The task can be embedded so students can stay focused while engaging in the process. The "write" component gives very specific guidance to support students whether they understand the problem or not. It also provides stems for students to consider any time they are engaged in solving a problem.

Think-Write-Pair-Share	
Think Think about the problem. INSERT TASK HERE	**Write** Write by doing one of the following: If you can solve, choose a strategy, and solve. If you cannot solve… ❏ Write all facts you know about the problem ❏ Write anything you know related to the concept addressed in the problem ❏ Write questions you have about the problem
Workspace	
Pair Pair with a partner and take turns discussing your strategies and solutions. Use this space to record strategies that were different from yours.	**Share** Share various strategies and solutions with the group. Use this space to record strategies that were different from those of you and your partner.

**See Section "Emphasis on Process over Solution" for Sample Lesson Plan using this tool.

Building Communication Skills While Attending to Precision

The following problem-solving process and graphic organizer were introduced in *Strategies for Common Core Mathematics: Implementing the Standards for Mathematical Practice* (Texas & Jones, 2013) and can be used to assist students in making sense of problems **(Standard of Mathematical Practice #1)** as well as decontextualizing and contextualizing word problems **(Standard of Mathematical Practice #2)**. The process also requires students to construct viable arguments **(Standard of Mathematical Practice #3)** as they formulate their own ideas about the meaning of the problem and make predictions about the outcome. Once a solution is obtained, students compare to the prediction to determine the reasonableness of the solution. By giving students explicit steps to unpack the problem, they begin the process with minimal to no teacher guidance and complete the initial steps. This eliminates the blank piece of paper or the famous, "I don't know" answer. Using a consistent process over time with students will assist them in becoming better problem solvers. While this process may not always "fit" every problem, it does help students develop a systematic approach to finding the "entry point" into various tasks.

The process is like the Three Reads strategy in that it asks students to read the problem more than once. The first time they read it in its entirety to understand the context of the problem. Steps 1 and 2 then ask students to reread specific sentences as they decode the text and make sense of the problem. See below for an explanation and how the graphic organizer is used to capture the process.

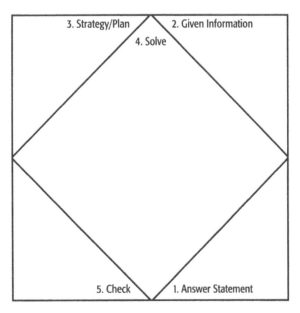

Authentic Opportunities for Writing about Math in Early Childhood

> This organizer can be enlarged and copied onto paper for students. It can also be created by folding a piece of paper at each end as if making a paper airplane. Once opened, it will be partitioned as above.

For an interdisciplinary version, see our book *Strategic Journeys for Building Logical Reasoning: Activities Across the Content Areas* (Jones & Texas, 2017).

1. **Answer Statement**
    a. The question usually appears as the last sentence of the problem. Students can cover the other information and focus on the last line to determine what the problem is asking. (If the question is not here, students can check each preceding line until it is found.)
    b. Students write the question as an answer statement and leave a blank for the solution. Translating from a question to an answer statement can be challenging for some students. Practicing verbally asking and answering the question can assist in this process.
    c. Remind students to include the appropriate units for the context of the problem.
    d. The answer statement is a critical component and should be practiced even when not using this organizer. It ensures students understand the question being asked as well as guarantees they will answer the question posed if the problem contains multiple steps. Developing this habit promotes its transfer to testing situations and is particularly important when answering constructed response questions.

2. **Given Information**
    a. Students use the same process of viewing each sentence separately, covering everything else.
    b. Students determine and record relevant information from the problem.

3. **Strategy/Plan**
    a. Students use this space to state additional ideas they have about the problem, such as other information they know about the problem, possible strategies for getting started, estimations for the solution, constraints, or predictions.
    b. This is the section that allows students to formulate their own ideas about the problem and provides a place for them to create their own meaning about what is being asked.

    c. Determining an estimate also provides a context for checking for reasonableness of the solution.
    d. This step also allows students to become strategic problem solvers rather than impulsive ones by requiring them to consider the various strategies available and then determine which might be the most efficient to use in the given situation.
    e. Many students are not versatile in the various problem-solving strategies available. Creating a Strategy Wall can be useful to build the students toolkit. See p. 26 for more information on Strategy Walls.
4. **Solve**
    a. Students select a strategy (translate verbal statements into mathematical statements, draw a picture, make a table, etc.) and solve.
    b. Students can compare their solution to the estimation to determine the reasonableness of their answer.
5. **Check**
    a. Students check their answers by substitution or by using another method to justify.
    b. This is also a good time to strategically partner students who used different strategies. Students can coach each other in the use of their strategy.
    c. Once the answer has been checked, students write the answer in the blank from Step 1.

Emphasis on Process over Solution

The purpose of any problem-solving process is to encourage students to think about the problem before impulsively jumping ahead to solving. It also encourages them to read and understand before assuming what is expected. To reinforce this point, students can be given a set of problems in which they are asked to complete the initial steps but not to solve. This allows the focus to be on making sense of the problem and planning before executing. If on a teaching team, this assignment can be completed in the ELA classroom since it involves decoding text and pre-writing skills. Once students have completed these initial steps, take away the problem set and have students complete the process by solving, checking, and answering the question. By not having access to the original problems, this will serve as an assessment of the initial steps. If students can complete the work, then the information gathered is sufficient. If not, it reveals key components that were overlooked.

Update and New Information

Since the publication of *Strategies for Common Core Mathematics: Implementing the Standards for Mathematical Practice* (Texas & Jones, 2013), many teachers have asked why the graphic organizer begins at the bottom right rather than the top left. There are two reasons it is organized in this manner. The first was in response to how the brain works when asked to attend. To focus and not just mindlessly record answers in a familiar sequence/order, the brain must consciously engage with the organizer and therefore students are more intentional with the process. The second reason was addressing when the process was internalized and the tool no longer needed. Most mathematics problems begin to be solved at the top left of the problem and then worked down to the bottom right where the solution usually is completed. This organizer begins with the end in mind (bottom right) and then comes full circle with the final answer.

The problem-solving process and the graphic organizer can be adapted to meet the needs of teachers and students and even eliminated as an organizer for students who internalize the process and no longer need the scaffold. Below is an example of a graphic organizer that was modified from the original. The first table contains scaffolds where there is a list of possible concepts/strategies for students to select as they build their toolkit. NOTE: The choices given here are general for illustration purposes and would be intentionally crafted for the specific unit in which it was being utilized. The second table has the supports removed.

Problem-Solving Process (Scaffolded)

The problem is asking me to... Answer statement:	I know...
Topic/Concept this is related to... Ex. Place value Performing mathematical operations Area/perimeter Fractions Other...	**Strategy for solving...** Draw a picture Guess and check Work backwards Use the standard algorithm Etc...
Solve (show work here)	
This solution means...	

Authentic Opportunities for Writing about Math in Early Childhood

Problem-Solving Process

The problem is asking me to...	I know...
Answer statement:	
Topic/concept this is related to...	Strategy for solving...

Solve (show work here)

This solution means...

Purposeful Writing: Intentional Design

Questioning: A Tool for Promoting Communication

As discussed in Section "Building Communication Skills While Attending to Precision" of our second series, "Strategic Journeys for Building Logical Reasoning," there are opportunities for questioning students while working through the problem-solving process.

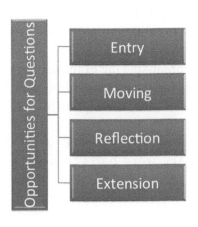

Entry: questions for students having difficulty getting started (Steps 1 and 2)
Moving: questions for places where students could get stuck (Steps 2–4)
Reflection: questions for students to use for metacognition after completing the problem/issue (Steps 4 and 5)
Extension: questions for students to engage in higher-order thinking skills with respect to the same concept and/or problem (after completing Step 5 and returning to Step 1)

These opportunities allow teachers to develop task-specific questions that can be used to support students as they are working through the process. In using these with students, it was noted that the first two opportunities occurred while students were in the middle of the process and the last two were once the process had been completed. Therefore, rather than viewing as four opportunities, they were condensed into two – I'm stuck and I'm done.

Four Question Types

1. Entry Questions
2. Moving Questions
3. Reflection Questions
4. Extension Questions

Task-specific questions can be generated and provided to students as needed, or they can be taught to access them on their own. If stuck, they can retrieve the appropriate questions that will allow them to move forward. For students who finish early, the done questions will be used to have them go deeper with the task rather than be assigned additional work, which is oftentimes seen as busy work.

See Section "Emphasis on Process over Solution" **for examples.**

Establishing Authentic Reasons for Writing

Incorporating literacy across the curriculum has long been an emphasis in mathematical classrooms. Initially, this involved having students put into words how they solved the problem alongside the mathematical steps. Fortunately, the redundancy of this request was soon realized. The mathematics itself clearly articulated what students did to solve the problem. Therefore, students were asked to write about their thinking rather than what they did. For example, if solving an equation, students were asked to write about the properties of equality used to explain why they did rather than what they did.

Section "Using with a Rich Task" introduces ten different strategies that can be used to provide authentic opportunities for students to write about math. Explanations of the strategies as well as content-specific examples have been provided to make these ready-to-use in the classroom. In addition, each provides the opportunity for differentiation. Below is a brief description of each:

Visual Prompts: pictures and images to initiate thoughts and discussions
Compare and Contrast: academic vocabulary word pairs to deepen understanding
The Answer Is…: giving an answer for which there could be multiple questions posed
Topical Questions: set of questions whose stems promote mathematical discourse
Writing About: using word clouds with academic vocabulary to write about a specific topic
Journal Prompts: assortment of ideas to engage students in journaling about mathematics
Poetry/Prose: collection of ideas to engage linguistic learners in expressing mathematical thought
Cubing/Think Dots: activities for independent learning
Question Quilts: an alternative way to present questions and provide student agency
RAFT: creative writing opportunity
Always, Sometimes, and Never: alternative way to view statements that promote critical thinking

PART TWO

Writing Prompts

CHAPTER 2

Visual Prompts

The visual prompts given here are actual photographs taken by the authors. These are different from what is known as "visual mathematics" which usually references the various visual representations in mathematics. The picture prompts harken back to *Mister Rogers' Neighborhood* Picture segments. These pictures help our students see the mathematics that is all around them. They also offer opportunities for students to engage in authentic mathematical communication.

The following collection of photographs can be used as journal prompts, discussion starters, bell ringers, or for centers, small groups, or learning stations. These pictures provide opportunities for students to engage in mathematics through looking at pictures of and from the world. As a starting point, students can free-write or discuss what they see and describe it. This could be facilitated much like the *Notice and Wonder* prompts that the National Council of Teachers of Mathematics has brought to the forefront in the past few years. Even for younger students who are just beginning to write, this provides an opportunity for them to put original thought to paper.

Students can name the colors they see, the geometric shapes they see used, the types of numbers they see, and the direction of an object such as the bear outside the Denver Convention Center standing up on the outside of the building looking in.

Authentic Opportunities for Writing about Math in Early Childhood

To begin discussions, you can provide one of the following prompts:

❏ What do you see?
❏ What colors do you recognize?
❏ What are some geometric shapes you see?
❏ How many _____ are there?
❏ How do you think math was used in this picture?
❏ What questions does the picture make you think about?
❏ What mathematical vocabulary could you use to describe the picture?
❏ Do you see any patterns in the picture? If so, describe the pattern.
❏ Where might you have seen something similar to what this picture is showing?
❏ In which picture is the bear outside looking in, the left or the right?
❏ Use directional/positional words to describe _____.
❏ Estimate how many _____ you think are shown. How did you think about that?

Take your own pictures of things in your town or school. Look for open-source images on the internet. Remember that visual prompts offer all students a voice and provide an opportunity for most students to enter the conversation and make mathematical connections.

Visual Prompts

Authentic Opportunities for Writing about Math in Early Childhood

Visual Prompts

Authentic Opportunities for Writing about Math in Early Childhood

Visual Prompts

Authentic Opportunities for Writing about Math in Early Childhood

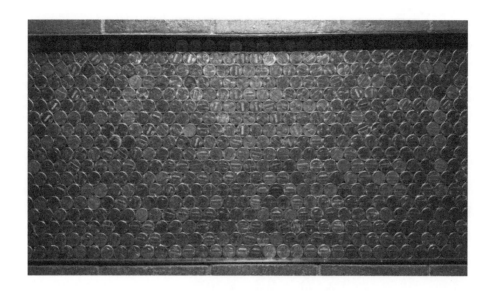

CHAPTER 3

Compare and Contrast

Writing math is typically a challenge for students. As discussed in our second series, "Strategic Journeys for Building Logical Reasoning," using the Mathematician's Notebook

> Can change the way you teach as well as how your students learn and experience their content. The notebook becomes a dynamic place where language, data, and logical reasoning experiences operate jointly to form meaning for the student" (Jones). A Mathematician's Notebook helps students create an organized space for demonstrating their learning process. The notebook serves as a formative instructional tool as well as a portfolio of the students' learning experiences.
>
> *(Jones & Texas, p. 14)*

Whether you are using a Mathematician's Notebook, an interactive notebook, or some other method of students chronicling their journey, all students need to be writing about math daily using paper and a writing implement.

Two of the main components of the Mathematician's Notebook are the glossary and the journal. Vocabulary is one of the foundations for developing an understanding of any subject area, and mathematics is no exception. Students need many opportunities to use their vocabulary in their daily work. Having students develop a glossary and reference the glossary as they progress through the year provides a resource for the students to use in their current mathematics course as well as future courses. Additional opportunities for students to engage with their academic vocabulary are vital for students to develop the deep understanding needed for success.

One such opportunity is the Compare and Contrast activity. Students can simply make a T-chart on their paper. They write the word pair (or three columns if using three words), one word at the top of each column. Students then compare the words by listing the ways they are alike and different. They write their ideas in the columns below each word pair. They conclude by writing a summary sentence about their ideas. If time, students can complete additional pairs. There is a graphic organizer provided if desired to use. It is set up so when copied it can be cut into half and used with two students.

Core mathematics vocabulary is provided to support work with emergent readers. These words and suggested activities offer opportunities for teachers to intertwine support for their students' language development with current mathematical topics being studied. These word pairs, as well as the word lists provided in the Writing About…section, can provide students with a beginning word bank. Pick and choose those word pairs that could be more accessible to your students. Manipulatives can be used with some of the word pairs. Pattern blocks, mini solids, and base ten manipulatives as well as any seasonal counters are always good options. Students can also incorporate gestures, drawings, and models as they reason and communicate mathematics.

Interactive Word Walls and Strategy Walls

Ideally, the vocabulary used in this activity would already be displayed on a word wall of key terms that have been discussed throughout the instructional unit. A strategic way to make a word wall more interactive would be to use words from the wall for this activity. Assign students the words or allow student choice, which would reveal how students are making sense of the relationships between the concepts. Once the activity is complete, have students display their work on the wall alongside the words.

To reinforce the idea of students building a toolkit of strategies that can be used when problem solving, a strategy wall is a helpful anchor chart. Using words from the additional lists below (create a list, create a table, and create a graph, draw a picture and draw a diagram, educated guess and random guess, eliminate possibilities and solve a simpler problem, formula and function, look for a pattern and use a formula, work backwards and work forwards, write an equation or inequality and model with manipulatives), build a strategy wall at the conclusion of the activity by displaying the words (problem-solving strategies) and student responses.

Compare and Contrast

A Beginning List of Word Pairs

Topic 1: Number and Quantity

Alike and different
Count on and count back
Equal and not equal
Greater than and less than
More than and fewer than
Number and numeral
Odd and even
Rod and flat
Skip counting and counting
Unit cube and rod

Topic 2: Algebraic Reasoning

Add and subtract
Addition and subtraction
Commutative property of addition and associative property of addition
Compose quantities and decompose quantities
Difference and sum
Doubles and doubles plus one
Doubles minus one and near doubles
Plus and Minus
Take away and put together

Topic 3: Geometric Reasoning/Measurement and Units

2D and 3D
A.M and P.M.
Analog clock and digital clock
Circle and hexagon
Closed shape and open shape
Compose geometric shapes and decompose geometric shapes
Cylinder and cone
Day and night
Face and side
Flat and curved
Heavier and lighter
Hour and half hour
Hour and minute
Hour hand and minute hand
Inch and centimeter

Authentic Opportunities for Writing about Math in Early Childhood

Inch and foot
Larger and smaller
Longer and shorter
Meter stick and yard stick
Pyramid and prism
Rectangle and square
Roll and stack
Shape and solid
Side and vertex
Slide vs stack
Tall and short
Trapezoid and rhombus
Triangle and rectangle
Whole and part of

Topic 4: Data Analysis, Probability, and Statistics

Bar graph and pictograph
Number line and line plot
Row and column
Vertical bar graph and horizontal bar graph

Additional Lists

Above and below
Beginning and end
Behind and in front of
Beside and between
First and last
Front and back
Left and right
True and false
Up and down

Compare and Contrast

Compare & Contrast

Choose a word pair. Write each word pair in the boxes below. Compare the words by listing the ways they are alike and different. Write your ideas in the columns below each word pair. Write a summary sentence about your ideas.

Word pair	
Compare:	
Contrast:	

Summary sentence(s):

Compare & Contrast

Choose a word pair. Write each word pair in the boxes below. Compare the words by listing the ways they are alike and different. Write your ideas in the columns below each word pair. Write a summary sentence about your ideas.

Word pair	
Compare:	
Contrast:	

Summary sentence(s):

CHAPTER 4

The Answer Is...

Students benefit from open-ended questions where there is possibly more than one correct response. This writing strategy allows students the opportunity to think beyond just procedural solving to get "the" answer. In some cases, the context is set up and given for the students. These questions will offer you as well as your students' insight into how they think about mathematics. Open-ended questions also encourage a growth mindset.

Students choose a card from "The Answer Is..." set to write about or to discuss, depending upon the level of the student. You can assign cards based upon students' individual needs. Students read the setup, if one is provided, then, they create a contextual problem for which the solution would be the answer given. This writing/discussion activity can be easily differentiated by setting parameters for students. The contextual problem can be a single step, or it may be multiple steps. The task may require a specific operation or include quantities within specific parameters. For example, kindergarten students may be working within ten, but in later second grade, students might be working with three-digit quantities. Students could also be required to provide at least two different possibilities for a context where the solution is the answer. Drawings, illustrations, and labels might also be needed for a complete response.

Note that some of the number and quantity tasks are focused on estimation rather than computation. These tasks could be put on index cards and used in a center or learning station. Students can even be encouraged to make their own "Answer is..." tasks.

The Answer Is...

Topic 1: Number and Quantity	
The answer is seven acorns. What could the question be?	The answer is ten. What could the question be?
The answer is 20 pumpkins. What could the question be?	The answer is more than **nine** What could the question be?

Authentic Opportunities for Writing about Math in Early Childhood

The answer is 25 acorns.

What could the question be?

The answer is 157 pumpkins.

What could the question be?

The answer is 500 acorns.

What could the question be?

The answer is 1000.

What could the question be?

The Answer Is...

Topic 2: Algebraic Reasoning	
The answer is three fish. What could the question be?	The answer is **four legs.** What could the question be?
The answer is What could the question be?	The answer is 0 + 5 What could the question be?

Authentic Opportunities for Writing about Math in Early Childhood

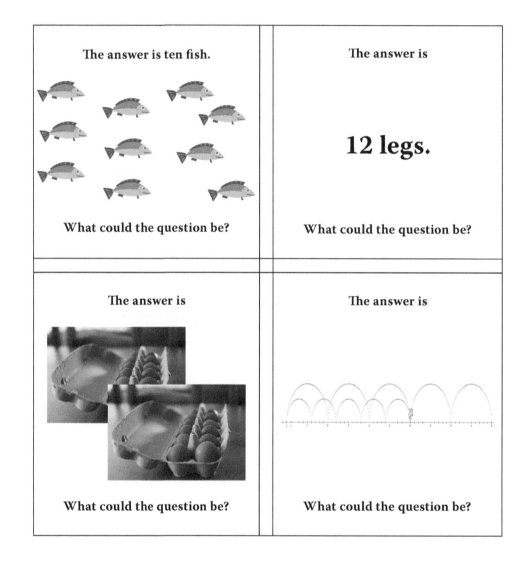

The Answer Is...

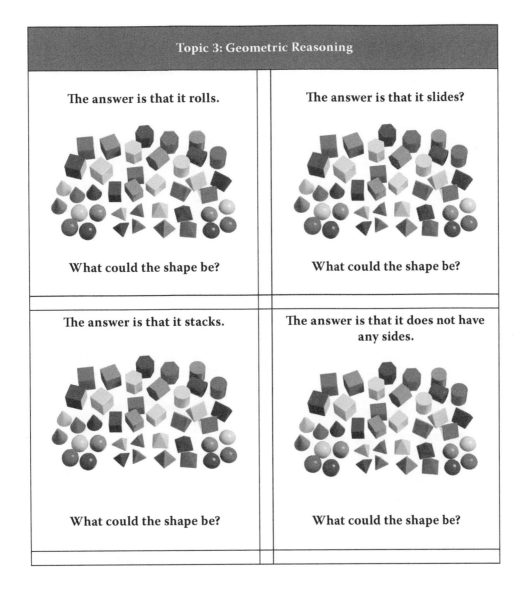

35

Authentic Opportunities for Writing about Math in Early Childhood

The answer is that it has four sides.  What could the shape be?	The answer is that it has more than four sides.  What could the shape be?
The answer is that it does not have any edges.  What could the shape be?	The answer is that one part of the object looks like a circle. 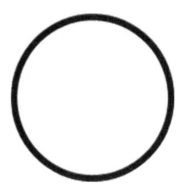 What could the object be?

The Answer Is...

Topic 4: Measurement and Data

The answer is 13 inches.

What could the question be?

The answer is $100.

What could the question be?

The answer is

What could the question be?

The answer is

What could the question be?

37

CHAPTER 5

Topical Questions

Questions are tools in teachers' toolbox and should be used as chisels to promote student thinking rather than pliers for answer-getting. Teachers practice, refine, and hone their questioning skills as they engage with students daily. Students can only provide a depth of answer based upon the quality of the question(s) asked.

As discussed in Section "Topic 3: Geometric Reasoning and Measurement" of our second series, "Strategic Journeys for Building Logical Reasoning," there are opportunities for questioning students that naturally exist when students are working through a task or activity. And the questions need to be carefully crafted so the mathematical discourse is not shut down. Using a question stem such as "is," "do," or "could" allows students the opportunity to simply answer yes or no and then they are done. If, however, you use a stem such as "how," "what," "when," or "where," along with others listed on our Q-Pyramid overlay (Jones & Texas, p. 92), you have opened the conversation and students must engage in mathematical discourse. This use of a more inquiring form of response encourages students to justify or explain their responses, whether they be correct or incorrect.

As you hone your questioning technique, be aware that one of the most important parts of questioning is how you respond to students. You should respond to your students in a manner that supports their thinking as it reveals to you what and how they are thinking. Wait time is vital as a quick response can often shut down the individual's or rest of the class's thinking and/or reflection on what is being said. Asking students to explain why, or to further discuss how they thought about something may at first be a struggle with students, but if it becomes a consistent part of your questioning, students will eventually accept it.

The questions provided in this section are not universal, but rather nuanced to the topic they reference. By design, they require communicating the answer more fully and are perfect for encouraging students to write about math. In the following, the topical examples start out with topics usually encountered in the entry level to upper elementary and extend to topics typically encountered in the upper elementary year. You can pick and choose based upon your grade level and the needs of your students. The suggested questions should generate additional questions as they are being discussed. Question prompts can be used to lead a discussion in class, or, in the later grades, as students have developed into more fluent readers, they can be run off and given to students as writing prompts, bell ringers, reflections, etc.

Note: For example, the first question, "What is a number?" is foundational to beginning mathematicians developing an understanding of quantities and their representations. Be sure discussions include the fact that a number represents a quantity of objects. The number (word) or numeral (symbol) answers the question, "How many?"

Authentic Opportunities for Writing about Math in Early Childhood

Topic 1: Number and Quantity

General

- ❏ What is a number?
- ❏ What is the connection between number and quantity?
- ❏ How does comparing quantities describe the relationship between the quantities?
- ❏ How can quantities be represented using symbols, words, pictures, and tools?
- ❏ What is a fraction?

Subtopic Specific

- ❏ **Counting**
  - ❏ What does it mean to count forward?
  - ❏ What does it mean to count backwards?
  - ❏ How does counting by five help you read a clock face?
  - ❏ What pattern(s) on the hundred chart help you count by fives? By tens?
  - ❏ How is skip counting represented on the number line?
  - ❏ What pattern(s) do you see when skip counting on a hundred chart by _____?
  - ❏ Why is zero not the first number when we count?
  - ❏ What pattern(s) do you see with the names of numbers that are greater than ten? Twenty? One hundred?
  - ❏ How do ordinal words relate to counting?

- ❏ **Making Sets**
  - ❏ How do you know when two sets are equal?
  - ❏ How can manipulatives be used to represent sets?
  - ❏ How can manipulatives be used to solve problems about sets?
  - ❏ How can sets be represented by a number sentence?
  - ❏ How are the arrangement of the objects being counted and the counting of the objects related?

- ❏ **Base Ten**
  - ❏ When making quantities larger than ten, what counting pattern(s) can be used?

Topical Questions

- ❏ What representations can you use when composing and decomposing quantities?
- ❏ Why is zero important in place value?
- ❏ What is place value?
- ❏ How does the position of a digit in a numeral affect the value of the digit?
- ❏ How does the expanded form of a quantity relate to the model of the quantity using base ten materials?

❏ **Comparisons**
  - ❏ What does it mean for a numeral to represent "one more?" "One less?"
  - ❏ What symbol is used to represent when two quantities are equal?
  - ❏ How do you know which symbol to use when comparing unequal quantities?
  - ❏ Do you find it easier to compare numerals, sets of objects, pictures of objects, or number words? Why?

Topic 2: Algebraic Reasoning

General

- What are some examples of situations that can be modeled by composing numbers or addition?
- What are some examples of situations that can be modeled by decomposing numbers or subtraction?
- Why is it helpful to recognize sums that make ten when working with addition and subtraction problems?
- Why is it helpful to recognize sums that make 100 when working with addition and subtraction problems?
- How are addition and subtraction related to counting?

Subtopic Specific

- **Composing/Decomposing Numbers**
  - What does it mean to "put together" two quantities?
  - What does it mean to "take apart" two quantities?
  - What are the different ways to compose ten?
  - How can the base ten materials be used to compose and decompose quantities?
  - Do you find it easier to solve situations with the total unknown, an unknown addend, or both addends unknown? Why?

- **Operations with Numbers**
  - What is a fact family?
  - What pattern(s) can be used when adding and/or subtracting with multiples of ten? Multiples of 100?
  - What is a "sum"?
  - What is a "difference"?
  - Why does the order of the addends not matter when combining two quantities?
  - Why is order important when determining the difference between two quantities?
  - How can "making ten" be helpful when solving an addition situation?
  - How can "doubles" be helpful when solving an addition situation?
  - What is the relationship between addition and subtraction?
  - What is meant by "properties of operations?"

Topical Questions

- **Models/Equations**
  - What does the equal sign represent?
  - What is a number sentence?
  - Does it matter which side of the equal sign the addends are on when modeling an addition situation? Explain.
  - How does an equation describe the addition/subtraction situation being modeled?
  - When is it more beneficial to use an equation to model an addition or subtraction situation versus using a drawing or base ten materials?

Topic 3: Geometric Reasoning and Measurement

General

- How can geometry help us make sense of our world?
- What is measurement?
- Why do we need measurement?
- What are some ways to measure?
- How do you choose the appropriate tool when measuring?
- How do you choose an acceptable unit when measuring?
- Can all things be measured? Explain.
- How are measurements related to counting?
- What are some positional words? Describe what they mean.
- How can positional words be used to describe shapes?

Subtopic Specific

- **Shapes, 2D and 3D**
    - What shapes are represented by the pattern blocks?
    - What is a dimension?
    - What does it mean for a shape to be two-dimensional?
    - What does it mean for a shape to be three-dimensional?
    - How does drawing a two-dimensional shape differ from drawing a three-dimensional shape?
    - What are some real-life objects that can be modeled by geometric shapes?
    - What shapes can be composed to create a hot air balloon, for example?
    - What shapes can be composed to create an ice cream cone, for example?

- **Composing/Decomposing Shapes**
    - How can triangles be used to compose a rectangle?
    - Into what shapes can a trapezoid be decomposed?
    - How does repeated addition and partitioning shapes model multiplication?
    - How does partitioning shapes model fractions?
    - How can circles and rectangles be partitioned into equal shares?
    - What names can be given to the resulting equal shares?

Topical Questions

- **Attributes of Shapes**
    - What is an attribute?
    - What are some attributes of two-dimensional shapes?
    - What are some attributes of three-dimensional shapes?
    - How are shapes named and described? For example, a closed shape with three sides is called a triangle.
    - Which solids can be stacked?
    - Which solids can slide?
    - Which solids can roll?

- **Geometric Measurement (Length, Width, Height, Perimeter)**
    - What are some tools that are used for measuring objects?
    - How is a ruler related to a number line?
    - What is meant by length?
    - What is meant by width?
    - What is meant by height?
    - What is meant by perimeter?
    - What are some shapes that have equal sides?

- **Time**
    - Why do we tell time?
    - What are the units for time?
    - How does reading a clock face relate to skip counting?
    - What is the difference between the "time of day" and the "length of time?"
    - What are some differences between an analog clock and a digital clock?
    - What do the abbreviations "a.m." and "p.m." represent?
    - What is a time zone?

- **Money**
    - What are three coins that we use and what quantity does each represent?
    - What symbols are used when writing money amounts?
    - What are at least three different ways to make one dollar using coins?
    - Is the size of a coin related to the quantity it represents? Explain.

Topic 4: Data

General

- ❏ What is data?
- ❏ How can objects be sorted?
- ❏ How can objects be categorized?
- ❏ Why can color be used to sort geometric shapes but not be used as a defining attribute?
- ❏ What is a picture graph?
- ❏ What is a bar graph?
- ❏ How can picture graphs and bar graphs be used to represent data sets?
- ❏ How are graphs used to sort geometric shapes by measurement attributes?
- ❏ How do graphs make it easier to compare measurement attributes of geometric shapes?
- ❏ What are some questions that can be answered about a data set by using a graph?

CHAPTER 6

Writing About…

Writing About is a small group writing activity that can be used strategically to support students who struggle with writing, particularly language learners. Just because a student can verbally tell you something does not mean that they can write that same response and support it with evidence. Prior to this activity, you might invite the ELA teacher to visit the class and share what makes a good paragraph so common expectations can be set that support the work in ELA.

Begin by giving students two or three index cards or scraps of paper. Students are to study the word cloud and write one or two sentences about the topic using words they find in the word cloud. Each student shares their sentences with the group and together create a paragraph about the topic. The index cards allow students to sequence the sentences to build a thoughtful and complete paragraph. They combine similar sentences, check for an introduction, conclusion, etc. This provides an opportunity for students to practice building a paragraph about a topic. As students first work in a group of three or four, they can then begin to work with a smaller group or a partner. The activity can be extended later as an individual writing activity as students are developing stamina for writing. Be aware that not all students will progress at the same pace.

Adaptations and Extensions: For children who have not yet begun to write and/or are emerging readers this activity can be adapted in a couple of ways. First, students can simply practice identifying the words and discussing what they mean, even in a circle or large group time. They can color each word

Authentic Opportunities for Writing about Math in Early Childhood

as they identify it. Note: There are two versions of each topic that can be sued based upon the needs of your students.

Second, children can create sentences and say them or record them. The sentences can be written on sentence strip paper so the sentences can be manipulated to create a simple three sentence paragraph or description. The words can be highlighted or circled that are being used by the children. Children learn to write by writing original thoughts. This activity provides students an opportunity to practice forming letters in a meaningful context while thinking about mathematics. This activity does take time in the earlier grades, but the dividends are worth it.

Students can sort the words found in the word clouds and create a mapping. Students can work in small groups, pairs, or individually. Students need to be able to articulate their sorting/mapping rule. If students are doing a mapping, they can draw connectors, use string/yarn, or use something like WikkiStix™. If using WikkiStix™, be sure students are working on a piece of construction paper or scrap paper that will not matter if the sticky gets on it or not.

Once students sort their word set and show their connections, they need to write down their sorting rule in their Mathematician's Notebook. Once all groups are finished, students can do a Walk About Review where they observe the other groups' mapping/sorting and make notes about what they think their sorting rules were. Then, the whole group can come back together and discuss what they observed. Some questions that you might use to facilitate the discussion could include:

- ❏ What were the similarities you observed between the mappings?
- ❏ What were some differences?
- ❏ Were you able to identify the correct sorting rule for the other groups? Why or why not?

Suggested directions for the mapping/sort:

Study the words. Sort the words. Sort the sets of words that seem to go together. You may use your string/WikkiStix™ to show connections between the words. Explain your sorting rule fully. If directed, create a second sorting with a different rule.

Suggested directions for the Walk About Review:

As you walk about and review the other groups mappings, do not talk, look over the mapping and in your notebook identify what you think the groups' sorting/mapping rule is and why. You will have a set amount of time at each mapping, so use it wisely and efficiently.

Ideas for Display: Groups can create a graffiti board using chart paper to capture their paragraph. These group boards can then be put together to create a graffiti wall. The class could do a gallery walk to view what was developed, provide feedback, and/or reflect on the process.

Writing About...

Writing about...

Study the word cloud below. Create at least two statements about addition and subtraction using the key words you see in the word cloud. With your group, use your sentences to create a paragraph about addition and subtraction.

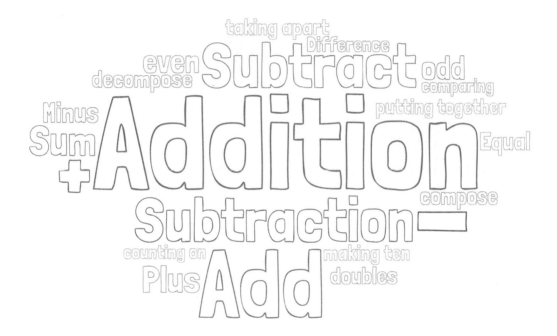

Authentic Opportunities for Writing about Math in Early Childhood

Writing about...

Study the word cloud below. Create at least two statements about addition and subtraction using the key words you see in the word cloud. With your group, use your sentences to create a paragraph about addition and subtraction.

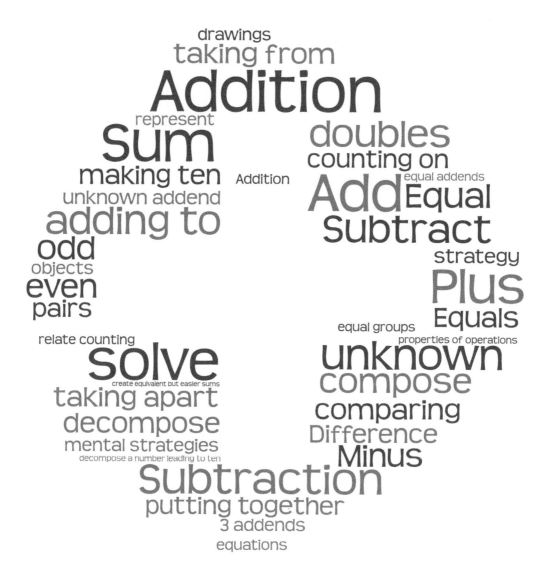

Writing About...

Writing about...

Study the word cloud below. Create at least two statements about counting using the key words you see in the word cloud. With your group, use your sentences to create a paragraph about counting.

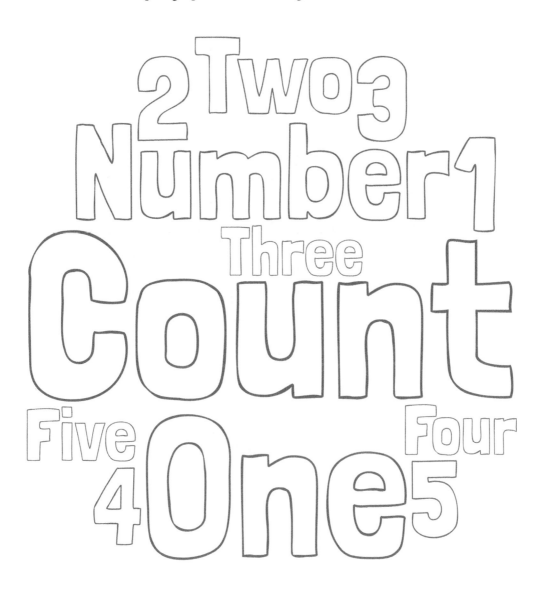

Authentic Opportunities for Writing about Math in Early Childhood

Writing about...

Study the word cloud below. Create at least two statements about counting using the key words you see in the word cloud. With your group, use your sentences to create a paragraph about counting.

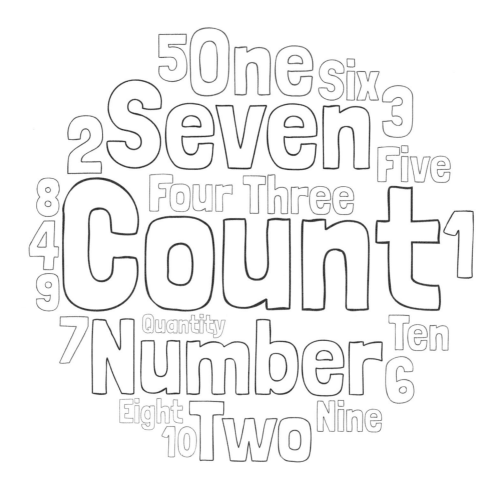

Writing About...

Writing about...

Study the word cloud below. Create at least two statements about counting using the key words you see in the word cloud. With your group, use your sentences to create a paragraph about counting.

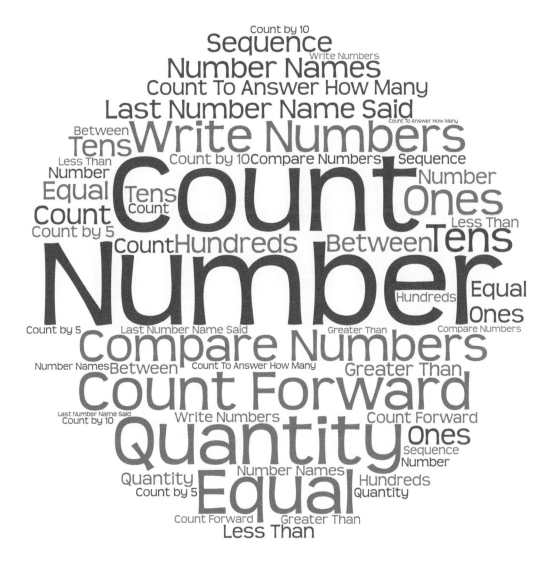

Authentic Opportunities for Writing about Math in Early Childhood

Writing about...

Study the word cloud below. Create at least two statements about geometry using the key words you see in the word cloud. With your group, use your sentences to create a paragraph about geometry.

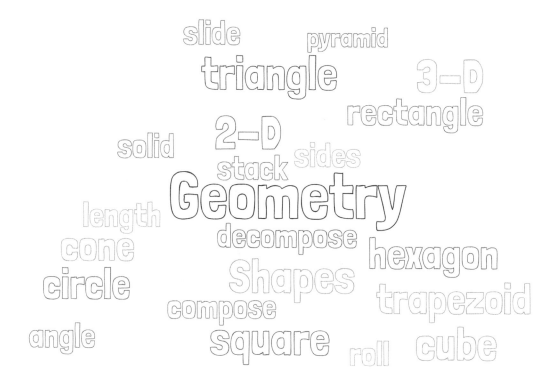

Writing about...

Study the word cloud below. Create at least two statements about geometry using the key words you see in the word cloud. With your group, use your sentences to create a paragraph about geometry.

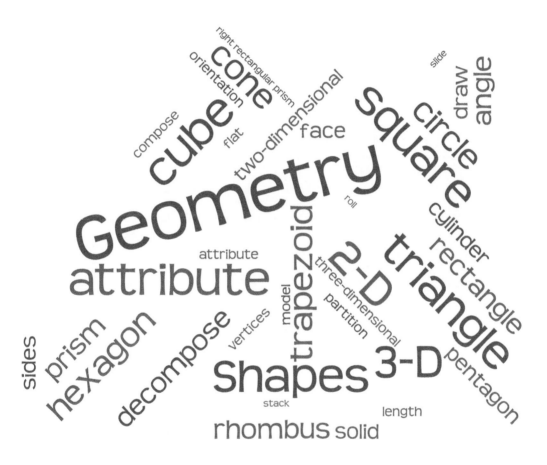

Authentic Opportunities for Writing about Math in Early Childhood

Writing about...

Study the word cloud below. Create at least two statements about money using the key words you see in the word cloud. With your group, use your sentences to create a paragraph about inequalities.

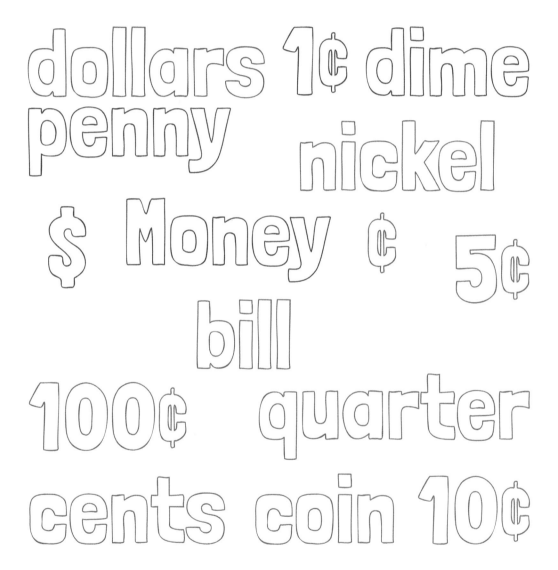

Writing About...

Writing about...

Study the word cloud below. Create at least two statements about money using the key words you see in the word cloud. With your group, use your sentences to create a paragraph about money.

Authentic Opportunities for Writing about Math in Early Childhood

Writing about...

Study the word cloud below. Create at least two statements about place value using the key words you see in the word cloud. With your group, use your sentences to create a paragraph about place value.

Writing About...

Writing about...

Study the word cloud below. Create at least two statements about place value using the key words you see in the word cloud. With your group, use your sentences to create a paragraph about ratios and place value.

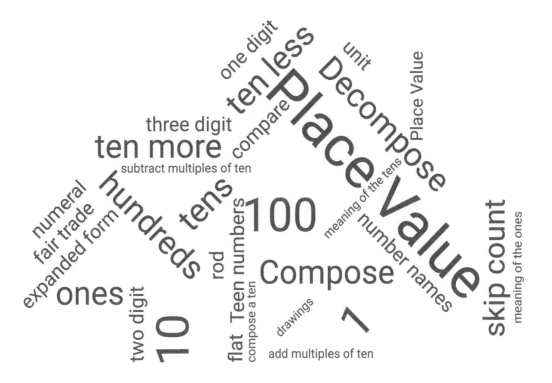

Writing about...

Study the word cloud below. Create at least two statements about positional words using the key words you see in the word cloud. With your group, use your sentences to create a paragraph about positional words.

Writing About...

Writing about...

Study the word cloud below. Create at least two statements about positional words using the key words you see in the word cloud. With your group, use your sentences to create a paragraph about positional words.

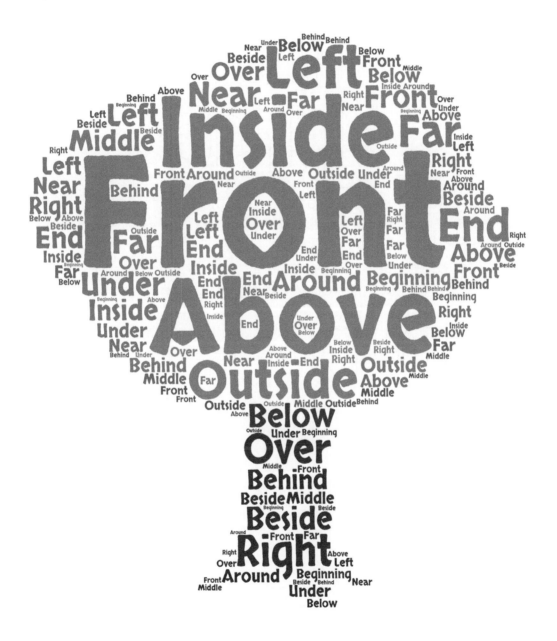

Authentic Opportunities for Writing about Math in Early Childhood

Writing about...

Study the word cloud below. Create at least two statements about time using the key words you see in the word cloud. With your group, use your sentences to create a paragraph about time.

Writing About...

Writing about...

Study the word cloud below. Create at least two statements about time using the key words you see in the word cloud. With your group, use your sentences to create a paragraph about time.

Topic 1: Number and Quantity

Count	**Number**
Quantity	**Compare**
Number names	**Numeral**

Writing About...

Sequence	**Ones**
Tens	**Hundreds**
Greater than	**Less than**

Authentic Opportunities for Writing about Math in Early Childhood

Skip count	**Count by five**
Count by 10	**Place value**
Teen numbers	**One digit**

Writing About...

Two digit	Three digit
Concrete model	**Drawing**

Authentic Opportunities for Writing about Math in Early Childhood

First	Second
Third	Fourth
Fifth	Sixth

Writing About...

Seventh	Eighth
Ninth	Tenth

Topic 2: Algebraic Reasoning

Addition	**Subtraction**
Unknown	**Add**
Subtract	**Putting together**

Writing About...

Adding to	Taking apart
Taking from	Solve
Decompose	Compose

Authentic Opportunities for Writing about Math in Early Childhood

Unknown addend	**Counting on**
Making ten	**Comparing**
Mental strategies	**Doubles**

Writing About...

Sum	**Difference**
Plus	**Minus**
Equal	**Doubles plus one**

Near doubles	**Fair trade**
Expanded form	

Topic 3: Geometric Reasoning

2-D	3-D
Angle	Attributes
Circle	Compose

Authentic Opportunities for Writing about Math in Early Childhood

Cone	**Cube**
Cylinder	**Decompose**
Draw	**Face**

Writing About...

Flat	**Geometry**
Hexagon	**Length**
Model	**Partition**

Authentic Opportunities for Writing about Math in Early Childhood

Pentagon	Prism
Pyramid	Rectangle
Rhombus	Roll

Writing About...

Shapes	Side
Slide	Solid
Square	Stack

Authentic Opportunities for Writing about Math in Early Childhood

Trapezoid	**Triangle**
Vertex	**Half of**
Fourth of	

Writing About...

Above	Around
Beginning	Behind
Below	Beside

End	Far Away
Close to	Front
Back	Inside

Writing About...

Outside	Left
Right	Middle
Over	Under

Topic 4: Measurement and Data

Analog clock	**Clock**
Clock face	**Count time**
Digital clock	**Count by five**

Writing About...

Count by 10	**Half hour**
Hour	**Minute**
Minute hand	**Tell time**

Authentic Opportunities for Writing about Math in Early Childhood

Time	Write time
Money	Cents
Dimes	Nickel

Writing About...

Penny	Quarter
Coin	Bill
$	¢

CHAPTER 7

Journal Prompts

As mentioned previously in the Compare and Contrast section, two of the main components of the Mathematician's Notebook are the glossary and the journal. Journals are a great way for students to keep track of their mathematical journey as well as giving you insight into how they think about math and how they have developed and grown over the course of the semester or year. Journals specifically designed for the primary grades with one-half of the page being blank for a picture and the lower half being writing lines are available. By second grade, most students could use the quadrille notebook, if desired. Journals can be a place where students engage with quotes, historical connections to topics, famous people related to the topic of study, misconceptions, and "What if?" scenarios.

Journal entries are best assessed separately from the rest of the Mathematician's Notebook. Even a basic *All, Most, Some, None* format works well. Journals are a place that you can dialogue with students about topics and engage with them in a different way. The authors never wrote directly on the student's notebook pages, as that was their own work, but wrote on sticky notes and attached it to the page(s) where comments were appropriate.

Following are some beginning suggestions for journal prompts that can be used throughout the year. Some are very focused around mathematical topics, and some are just for fun for students to allow their imagination to run wild. Hopefully these will serve as the basis for you to add many additional ideas of your own.

How I interact with the math around me: Students are prompted with the question, "What do they think math is?" Students can be given playdough

Journal Prompts

to form math objects, shapes, symbols, etc. Students can draw math objects, shapes, symbols, etc. Students who are writing original thought, or as an opportunity to write original thought, can be encouraged to write about what they think math is, or includes.

> ### WHAT IS MATH?
>
> **Think About**
> - Earliest remembrances of counting, learning about numbers
> - Pre-school work
> - Things you did at home with your parents/caregivers
> - What you see in the classroom that makes you think of math
> - What you see in your world that makes you think of math

Writing to Explain

Option 1: Students write or record an explanation for a student in their class who was absent the day they learned about/how to (insert topic/activity/procedure for the day here.)

Option 2: Students are assigned a mathematical topic such as counting, identifying shapes, or adding within 100. Then they complete the following.

- ❏ You are zero. Tell us everything we should know about you.
- ❏ You are a square. Tell us everything we should know about you.
- ❏ You are addition. Tell us everything we should know about you.
- ❏ You are the tens place in a number. Tell us everything we should know about you.
- ❏ You are a book. Describe yourself, including your measurements.

Creative Writing

For early kindergarten, students can discuss their replies to the prompts, record them, or draw their response. Students complete each of the following prompts using their imagination. Encourage students to just not write but to also use drawings and sketches and color as they complete the prompt.

- ❏ If I were a number, I would be _____ because....
- ❏ If I were a geometric shape, I would be _____ because....

Authentic Opportunities for Writing about Math in Early Childhood

- Draw something small. Now draw something large. Name the objects.
- You wake up tomorrow morning and find that circles no longer exist. What would change in your world if there were no circles?
- If I were a pattern, I would be _____ because....
- What I find the most challenging with _____(current topic) is.... Explain why.
- When I see a math problem with words, I feel _____ because....
- Choose one of your favorite characters from a book, a movie, tv show, etc., and describe how he/she might use mathematics in what he/she does.
- Just doodle mathematics until you fill an entire page. Share your doodles and discuss what you see in the doddles if anything.
- Fill a blank page with shapes. Color and decorate each shape differently. Identify and describe as many shapes as possible.
- Write a compliment to yourself for something you accomplished in math recently.
- Create a story using the numbers representing everyone's age in your family.
- Use a time number line to document a day in your life. Write a narrative describing your day.

Using Quotes

Option 1: Students write the quote, they write what it meant in the time it was written, and then how it would be applicable to them in math class today.

Example: *Arithmetic is being able to count up to twenty without taking off your shoes.* – Mickey Mouse

- Students copy the quote in their journal.
- They then write a couple of sentences about what they think Mickey meant.

Option 2: Students write a response to the author.

Option 3: Students write about what questions the quote prompts them to think about.

Option 4: The students describe what the quote means to them.

Below is a beginning list of quotes which span from historical to modern day, includes a diverse group of individuals, and cuts across disciplines to include the humanities as well as the sciences.

It's amazing what one can do when one doesn't know what one can't do. – Garfield the Cat[3]

Often, we put children in a box when if we let them think and reason and investigate on their own they surprise us with what they are capable of accomplishing.

Number is the within of all things. – Pythagoras[4]
Numbers are everywhere – students need many opportunities to describe their world through numbers.
Nature's Great Book is written in mathematical symbols. – Galileo Galilei[3]
What numbers and number patterns do we see in the world around us? Take a math nature walk and collect things that show "1," show "3," etc. Look for shapes in nature.
If you don't like the answer, ask a different question. – Dr. Larry Fleinhardt – NUMB3RS[5]
Having students create questions given an answer provides a more cognitively demanding experience for students as well as supports the development of deeper understandings of the mathematics being studied.
Mathematics is a game played according to certain simple rules with meaningless marks on paper. – David Hilbert[6]
Even primary mathematicians being to see the order and structure to mathematics.
Millions saw the apple fall, but Newton asked why. – Bernard Baruch[3]
Primary mathematicians need to learn to ask why!
The only way to learn mathematics is to do mathematics. – Paul Hamos[1]
Math: the only subject that counts. – Anonymous[1]
You never fail until you stop trying. – Albert Einstein[1]
Wherever there is a number, there is beauty. – Proclus[2]
Mathematics is the gate and key to science. – Roger Bacon[2]
But in my opinion, all things in nature occur mathematically. – Rene Descartes[2]
The important thing to remember about mathematics is not to be frightened – Richard Dawkins[2]

References

1. https://brighterly.com/blog/math-quotes-for-kids/
2. https://www.splashlearn.com/blog/brilliant-math-quotes-that-you-can-share-to-inspire-students/
3. https://www.goodreads.com
4. https://quotefancy.com
5. https://www.quotes.net
6. https://quotefancy.com

CHAPTER 8

Poetry/Prose

In the spirit of writing in response to a quote (described in the Journal Prompts), this section begins with one from JoAnne Growney's blog Intersections – Poetry with Mathematics

> Mathematical language can heighten the imagery of a poem; mathematical structure can deepen its effect.

The precision of language required of both disciplines makes the intersection of mathematics and poetry seem almost obvious. Providing the opportunity to see the connection allows students to explore this relationship while deepening their understanding of mathematics and writing skills.

Acrostic: explain – one word, expression, describing the main word…
   Example:

Circle

Curvy
Inside
Round
Colorful
Loopy
Empty

Math

Mysterious to some.
Ability required.
The language of our world.
Hurts my brain!

Beginning List of Terms: (see Word Lists in Compare and Contrast for Additional Terms)

Add
Subtract
Geometry (or any specific shape or term)
Count
Math
Numbers (or use any specific number set)
Ruler
Inch (or any measurement)
Time (or any measurement of time)

Fibonacci poem: Students can create their own "Fibonacci" poem where each line of the poem has the number of words as found in the sequence. This can be differentiated for students by having them use only the first three or four numbers found in the sequence or more if their writing skills allow. The topic of the poem can be of their choosing or can be assigned. The following poem models 1, 1, 2, 3, 5, 8.

Triangles

Triangles
Pointed
Three sides
And three angles
Walking from vertex to vertex
Around the corners, either obtuse, acute, or right.

Haiku: Has three lines, five syllables in first and third lines, and seven syllables in the second line
   Example:

Adding

Digits are combined
Sum of parts, parts of a whole
Can I now make ten

Pi poem: Is similar to the Fibonacci poem. Students write lines based on the digits in pi: 3.1415926535 8979323846 2643383279. This can be differentiated for students by having them use only the first three or four digits found in pi or more if their writing skills allow.

Students can also have fun with the topic as in the example below. Example:

Pi

My favorite pie –
cherry.
Flakey crust, tart flavor,
Yum!

Free verse/free write: There is no specific form, meter, or rhyme scheme. Students can have the freedom to write as they feel. Some suggestions are given below for types of free writes students may enjoy doing.

Cartoons

Commercial, Infographic, Public Service Announcement (PSA)
Free Verse Poem
Graphic novel, for example, The Adventures of Slope Boy
Historical Fiction
Math Carols and/or seasonal songs
Math words to a current song
Short Story

CHAPTER 9

Cubing and Think Dots

Cubing and think dots are two strategies for differentiation in the classroom. Traditionally students are given a cube with a variety of activities or tasks. Different cubes can contain different levels of information, tasks, and activities. Think dots work in a similar way. You can let students work through the six cards, or you can choose the parameters for your students. You can also create your own set of think dot cards using index cards and practice problems from your chosen curriculum.

For younger students, these activities can be introduced during the morning meeting, circle time, or in small group work. These activities can also be used to support intervention work with students. As students are introduced to the various activities and begin their work, encourage them to make statements explaining and supporting their reasoning and thinking. As students begin using written responses, they need to be able to use precise mathematical language and symbols in their written work as well as clearly articulate their thinking.

Preparation

Print the various "dice" and think dot cards on cardstock or heavy paper. Build the "dice" carefully, using the dotted lines as the folds. Tape or glue the edges together. Fold the think dot cards and tape together. These resources can also be used in a center or learning station. Some suggestions are given below.

Materials List:

- ❏ Three sets of think dot cards
- ❏ Number names tetrahedron to make
- ❏ Numerals tetrahedron to make
- ❏ Pips cube to make
- ❏ Numeral cube to make
- ❏ Number names cube to make
- ❏ Digits decahedron to make
- ❏ Number names decahedron to make
- ❏ Pips decahedron to make
- ❏ Teens numerals decahedron to make
- ❏ Teens number names decahedron to make
- ❏ Action cube 1 to make
- ❏ Action cube 2 to make
- ❏ Action tetrahedron to make
- ❏ Optional:
  - ❏ Various counters/manipulatives
  - ❏ Number cubes and/or dominoes (Note: You can purchase dice for games that are in various shapes beyond the basic cube and domino sets that show numbers greater than six.) Platonic solid sets of dice include the 4-sided tetrahedron, the 6-sided cube, the 8-sided octahedron, the 12-sided dodecahedron, and the 20-sided icosahedron. These allow students to work with larger quantities as they are ready for each of the various activities and tasks.

Tetrahedrons: The tetrahedrons are provided if some students need support working with the numbers one to four. They only offer the numerals 1–4 as well as the number names one to four. The action cube also covers the four basics for early work with numbers: adding, subtracting, comparing, and representing. Students can begin their work with the tetrahedrons by simply identifying the numerals and representing them with manipulatives, drawings, or gestures, such as clapping. Later, they can roll the tetrahedron twice or you can make two of them so students can perform basic addition and subtraction problems. Manipulatives, drawings, etc., can also be used here.

Cubes: There are two cubes provided, the numeral cube and the number names cube. They differ from a typical cube die by having the digits from 0 to 5 and the

Cubing and Think Dots

number names from zero to five on them. These can be used initially for most students. You can use a standard die to show the pip patterns and either leave the six pips as is, or you can cover it with a piece of tape to represent zero. The same types of activities that are described above can also be done with these cubes. They are also used with a couple of the think dot cards in the number and quantity beginning level. The action cubes are described below.

Decahedrons: The decahedrons are provided to provide for the digits 0–9 as well as the number names from zero to nine. There is a second set that covers both the numerals and number names for the teen numbers. Again, the basic activities first described with the tetrahedrons can be used here. These "dice" are used with the number and quantity levels 1 and 2 think dot cards.

Action Cube Options

Action Cube 1: (Used Primarily with the Beginning Level)

Count Forward: Students count forward from the number they rolled as many digits as you want them to. If using this in large group and/or with the teens "dice," the first student who rolls could count forward with other children following suit and see how far they can go without making a mistake.

Count Backward: Students count backward from the number they rolled as many digits as you want them to, or as they can. If using this in large group and/or with the teens "dice," the first student who rolls could count forward with other children following suit and see how far they can go without making a mistake.

One More: Students give the number that is one more than the one they rolled.

One Less: Students give the number that is one less than the one they rolled.

Double: Students give the number that is the double of the one they rolled.

Action Cube 2

Add: Students will combine the two numbers they roll on two rolls or on two dice. Students can be required to write the mathematical equation, represent their work with a drawing, or use manipulatives to explain their thinking. As students are ready, they can use three addends.

Subtract: Students will subtract the two numbers they roll on two rolls or on two dice. Students can be required to write the mathematical equation, represent their work with a drawing, or use manipulatives to explain their thinking.

Compare: Students will compare the two numbers they roll on two rolls or on two dice. Students can be required to write a mathematical sentence using the inequality symbols as they are ready, or they can represent their work with a drawing or use manipulatives to explain their thinking. The goal in comparing is for the students to be able to recognize the larger and the smaller quantity from just looking at the numeral.

Represent: Students will represent the number they roll Students can be required to write a mathematical sentence using the inequality symbols as they are ready, or they can represent their work with a drawing or use manipulatives to explain their thinking.

Make 10: Students will identify the quantity that "makes 10" with the number they roll.

Make 20: Students will identify the quantity that "makes 20" with the number they roll.

Think Dot Cards

Number and Quantity Beginning Level

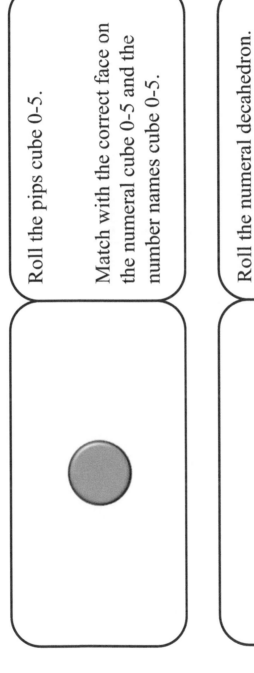

- Roll the pips cube 0-5.
- Match with the correct face on the numeral cube 0-5 and the number names cube 0-5.

- Roll the numeral decahedron.
- Match with the number names decahedron and the pips decahedron.

Authentic Opportunities for Writing about Math in Early Childhood

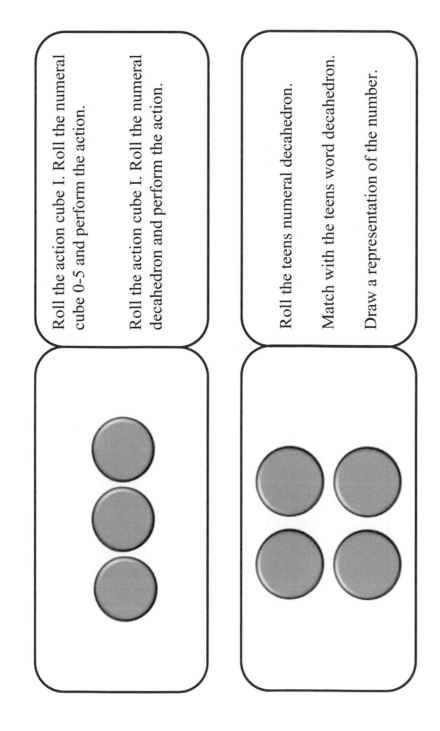

Cubing and Think Dots

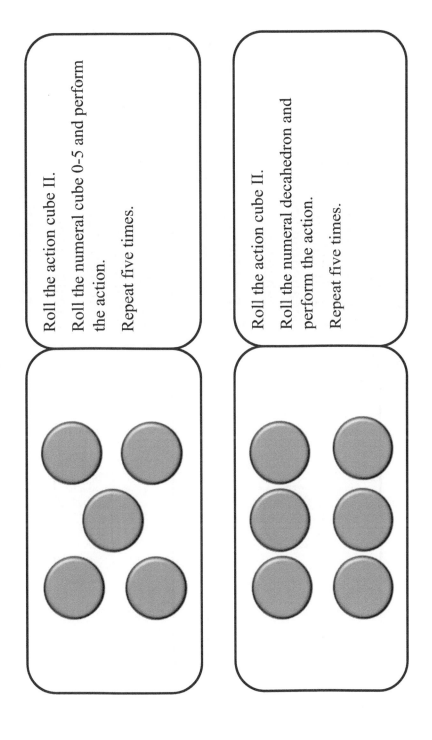

Roll the action cube II.

Roll the numeral cube 0-5 and perform the action.

Repeat five times.

Roll the action cube II.

Roll the numeral decahedron and perform the action.

Repeat five times.

Authentic Opportunities for Writing about Math in Early Childhood

Think Dot Cards

Number and Quantity Level 1

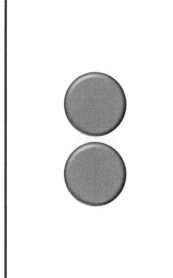

- Roll the numeral decahedron.
- Match with the number names decahedron and the pips decahedron.
- Repeat three times.

- Roll the teens numeral decahedron and match with the teen word decahedron.
- Create a representation.
- Repeat three times.

Cubing and Think Dots

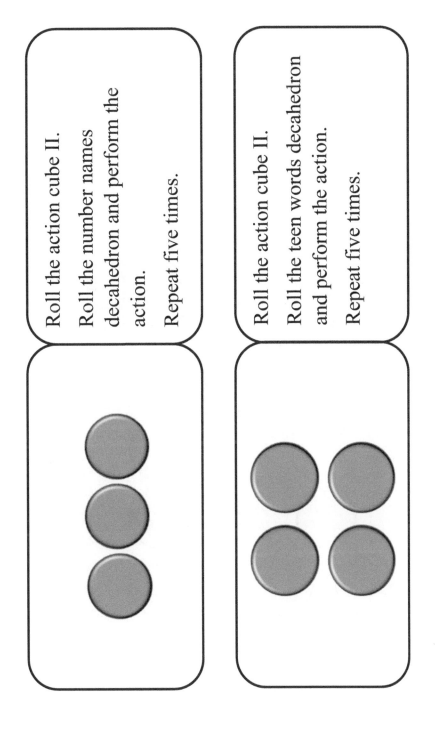

Roll the action cube II.

Roll the number names decahedron and perform the action.

Repeat five times.

Roll the action cube II.

Roll the teen words decahedron and perform the action.

Repeat five times.

Authentic Opportunities for Writing about Math in Early Childhood

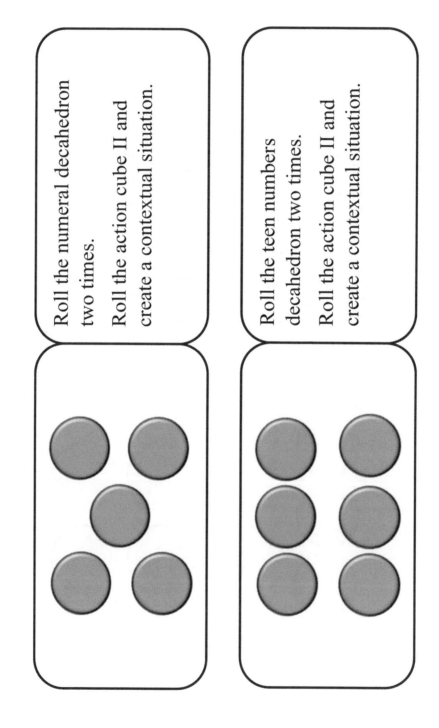

Cubing and Think Dots

Think Dot Cards

Number and Quantity Level 2

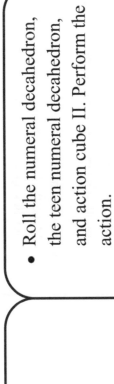

- Roll the numeral decahedron, the teen numeral decahedron, and action cube II. Perform the action.
- Repeat five times.

- Roll the numeral decahedron four times and create two 2-digit numbers.
- Roll the action tetrahedron and perform the action. Repeat five times.

105

Authentic Opportunities for Writing about Math in Early Childhood

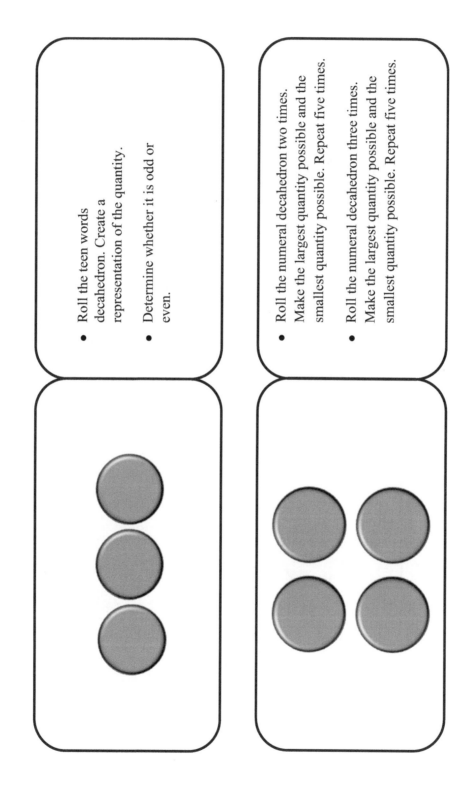

Cubing and Think Dots

- Roll the numeral decahedron three times. Make the largest quantity possible. Write the number name. Write in expanded form. Represent with base ten materials.
- Roll the numeral decahedron three times. Make the smallest quantity possible. Write the number name. Write in expanded form. Represent with base ten materials.

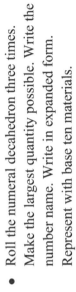

- Roll the numeral decahedron four times and create two 2-digit numbers. Create a 1-step contextual situation and solve showing your strategy.
- Roll the numeral decahedron four times and create two 2-digit numbers. Create a 2-step contextual situation and solve showing your strategy.

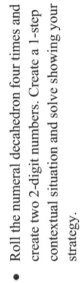

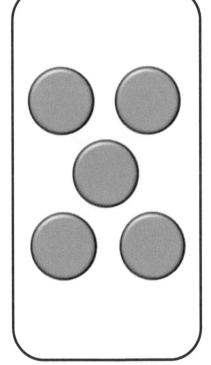

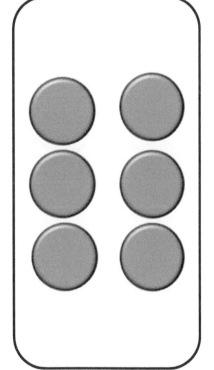

Authentic Opportunities for Writing about Math in Early Childhood

Number Names Tetrahedron

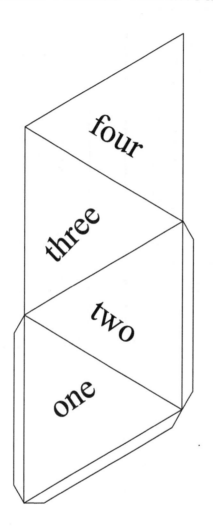

Numeral Tetrahedron

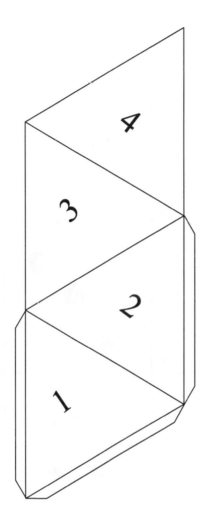

Authentic Opportunities for Writing about Math in Early Childhood

Pips Cube

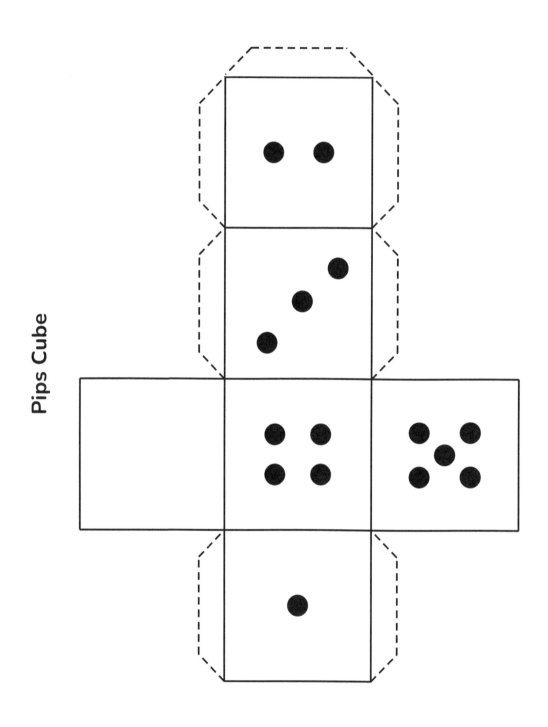

110

Cubing and Think Dots

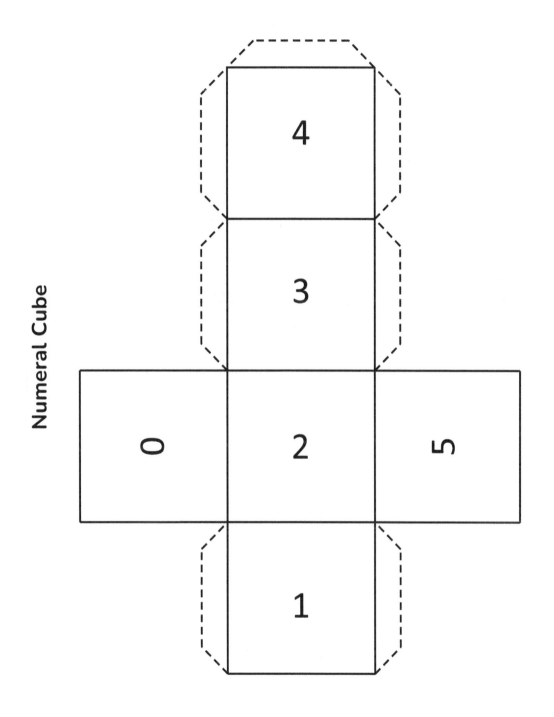

Authentic Opportunities for Writing about Math in Early Childhood

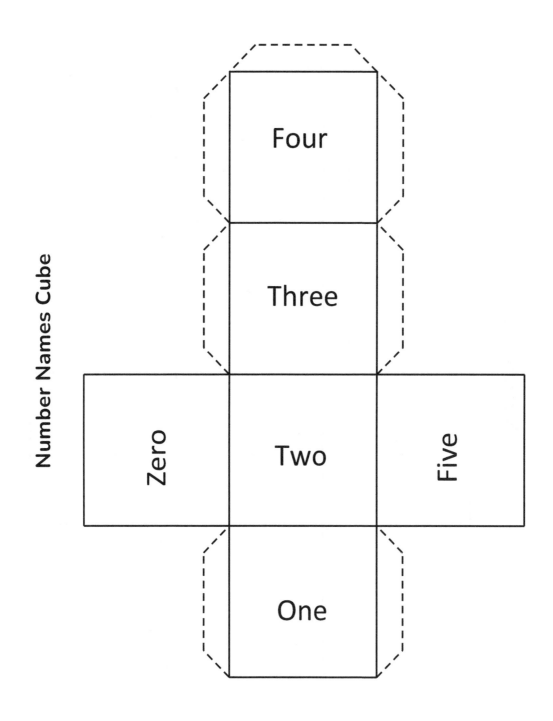

Number Names Cube

Numeral Decahedron

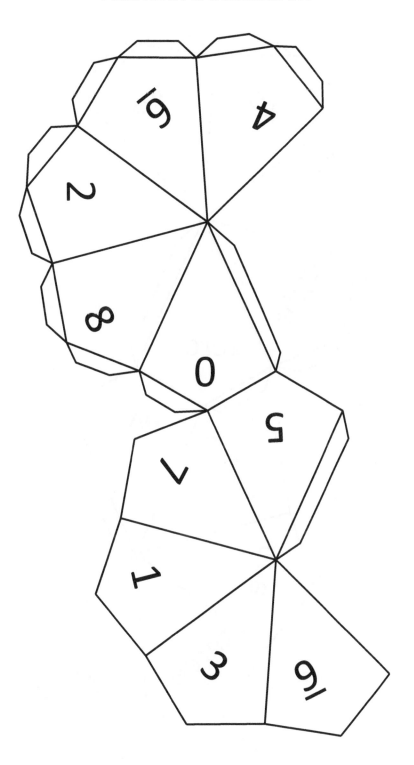

Cubing and Think Dots

Authentic Opportunities for Writing about Math in Early Childhood

Number Names Decahedron

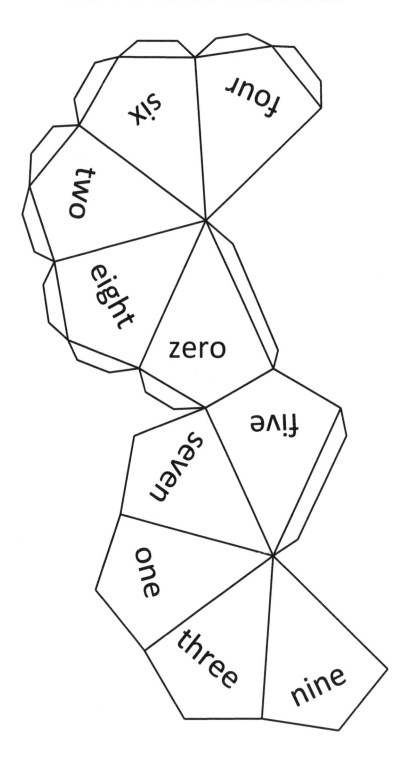

114

Decahedron Pips

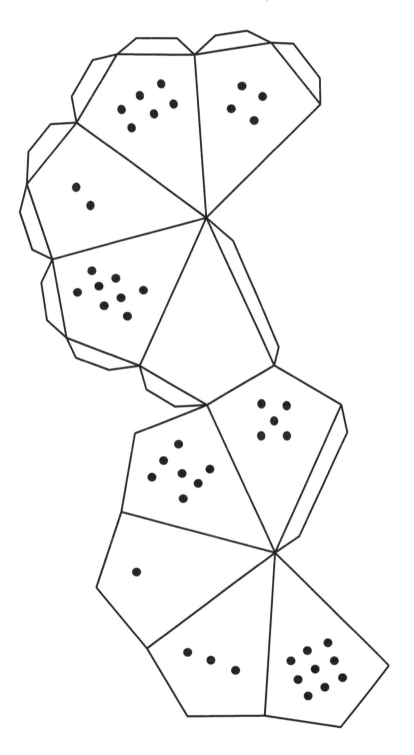

Authentic Opportunities for Writing about Math in Early Childhood

Teens Numeral Decahedron

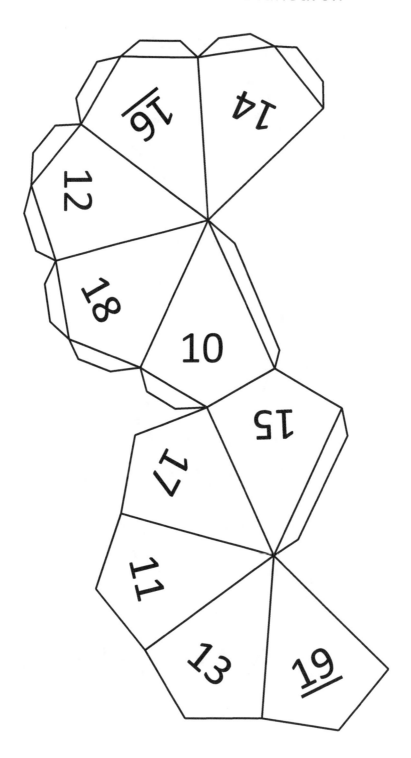

Cubing and Think Dots

Teens Number Names Decahedron

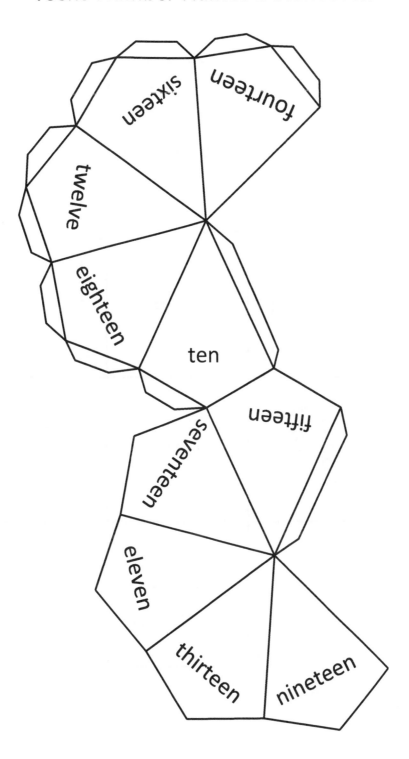

Authentic Opportunities for Writing about Math in Early Childhood

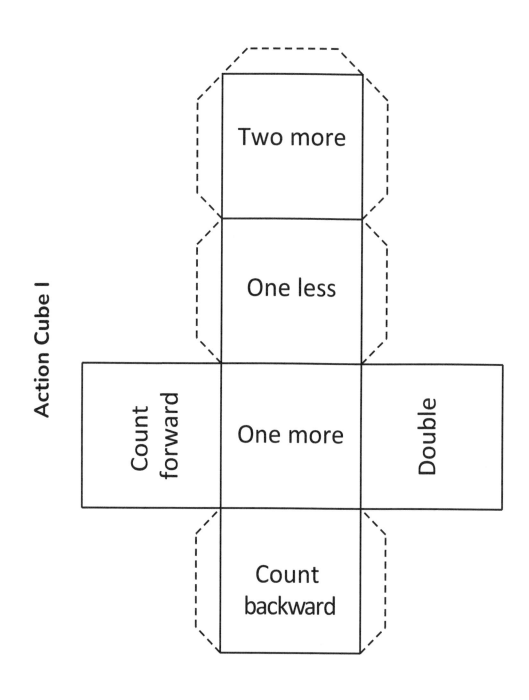

Action Cube I

118

Cubing and Think Dots

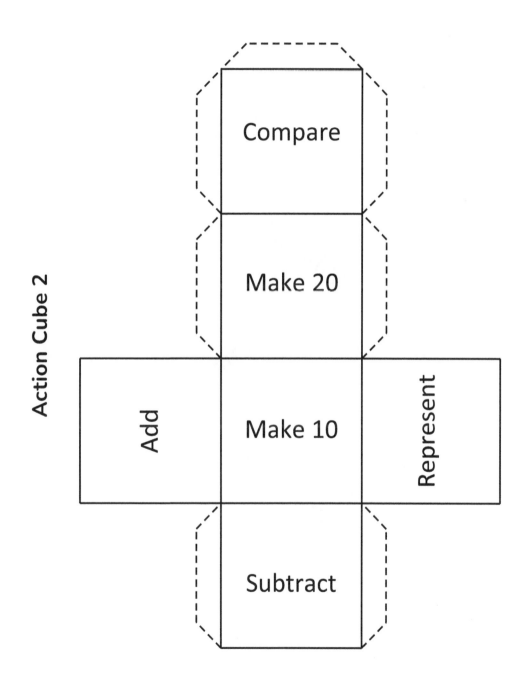

Action Cube 2

119

Action Tetrahedron

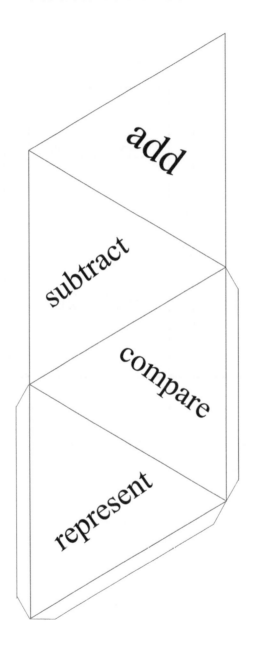

CHAPTER 10

RAFT

The RAFT writing strategy allows students to utilize their creativity within a mathematical context. Students are given a situation, along with a structure, which sets the parameters for the piece. If using as a formative assessment, the topic can be kept vague, as the bold text indicates, to determine whether students have internalized the key understanding(s) of the topic. If the intention is to ensure specific mathematical points are included in the piece, clearly state that expectation within the topic guidelines as the additional text demonstrates. This could be the form of a rubric.

Role	Audience	Format	Topic
One-digit number	10	Show and tell	**How to get to you.** Explain how to make ten based on your number.
Addition	Subtraction	Family grouping	**How we are related:** Discuss the fact family for a given pair of numbers.

Authentic Opportunities for Writing about Math in Early Childhood

Role	Audience	Format	Topic
Green triangle	Pattern blocks	Build a model	**How many?** Model how many green triangles it takes to compose each of the other pattern block shapes.
Box of toy vehicles	Student	Create a Venn	**Write a rule:** Sort the box of vehicles using a Venn and write your sorting rule.
Zero	Natural numbers	Song	**The role I play:** Discuss the role of zero in math
Sphere	Circle	Cartoon	**What's the difference?** Compare and contrast 2D and 3D shapes.
Nickel	Dime	Foldable	**It's not always as it seems:** Explain how the value of a coin is not based upon the size of coin.
Minute hand	Hour hand	Race stat sheet	**Race around the clock:** Describe the position of each of the clock's hands as they travel over an hour period of time.
Number line	Hundreds chart	Poster	**Skip on me:** Show skip counting patterns on both a number line and a hundreds chart and compare.

RAFT

Using RAFTs in primary grades can sometimes be challenging. One good way to incorporate the idea of RAFT is letting students create something tangible. Children love to color and draw. Pick a topic, for example, 2-D shapes. Have the students create a coloring book that next year's students can color and work in as they learn about the topic. A sample is given below as a suggested place to begin.

Shape	Attributes
	Name of shape: Color the shape Green. Color the **vertices** Blue. Put an "x" on each of the **sides**.

Role	What is the writer's role?	Yourself
Audience	Who will be reading the piece?	Next year's students
Format	What is the best way to present the information?	Coloring book
Topic	Who or what is the subject?	Color my math

123

Additional Format Ideas

Speech	Journal Entry	Script	Invitation
Public service announcement	Greeting card	Story board	Poem
Text message	Petition	Letter (apology, persuasive, thank you, complaint)	Commercial
Advertisement	Recipe	Campaign speech	Wanted poster
Itinerary	Personal ad	Nursery rhyme/riddle	Infographic
Editorial	Eulogy	Current event	Debate

Additional Samples

Role: Making Ten
Audience: Skip Counting
Format: Debate
Topic: My strategy is better than your strategy!

Role: Near Doubles
Audience: Doubles
Format: Partnership agreement (can use a table format)
Topic: How we work together: list the different doubles that can be used in the near doubles example

Role: Multiplication sign
Audience: Plus sign
Format: Phone Call
Topic: I'm a lot like you.

CHAPTER 11

Question Quilt

The question quilt can be a strategy for differentiation that allows for student agency as they read, or are read to, and think about questions and statements relating to a topic of study. Students decide if they agree or disagree with the statement(s) and/or answer the question(s) and discuss and/or write justification supporting their responses. Questions and statements can be framed to accommodate a variety of levels of learners. Some of the questions/statements are more geared to the lower grades in the grade band where others are more geared to the upper grades.

When using the questions with younger children, the quantities can be adjusted as needed. Manipulatives, hundred charts, number lines, etc., can be provided for student support as well. If using in large group, it is suggested that you make two copies. Cut one out and Velcro it on top of the other. You could also enlarge the quilt as well.

Sample directions: For students who are more independent readers, "Choose at least three questions or statements from the question quilt. Answer the question or decide if you agree or disagree with the statement. Justify your responses fully."

Authentic Opportunities for Writing about Math in Early Childhood

Question Quilt
Counting

- A friend is just learning about numbers. Explain to them counting on.
- Describe skip counting by….
- How are doubles and even numbers related?
- Discuss how counting by 1's is like counting by 10's.
- Discuss how counting by 10's is like counting by 100's.
- Tammy counts to 9. Counting back by one she says is 8.
- There are some toys in the toy box. How can you decide how many are in the toy box without counting?
- When I count, I begin with 0.
- Leslie says, "Odd numbers are every other number on the number line." Tammy says, "Even numbers are every other number on the number line." Who is correct?
- On a number line, the larger quantity is to the right.
- I have a stack of three books. My friend has a stack that is taller. How can I compare who has more books?

Question Quilt
Adding and Subtracting

- Jamal says that adding and subtracting are really the same thing.
- The difference is the answer when adding.
- Use base ten materials to represent a subtraction that needs to be regrouped.
- Discuss which tools you can use to help you model operating with numbers. Include how they help you.
- Explain different ways we can combine numbers to get the same result.
- Show at least three ways you can represent the following: Some fish swim into the pond, there are already 5 fish in the pond. Now there are 8 fish in the pond.
- Describe how you would estimate the sum of 46 and 85.
- When we put numbers together, we call it….
- Today you learned about adding and subtracting by 10. Discuss the patterns you noticed. Extend this to adding and subtracting by 100.
- 47 − 18 = 31

Authentic Opportunities for Writing about Math in Early Childhood

Question Quilt
Time and Money

- Sketch and label a clock for each of the following:
  - the time you get up in the morning,
  - the time you come to school, and
  - the time you go to bed.

- Create a time chart showing your typical school day.

- When making change from a dollar, counting up is a good strategy.

- How many minutes are in an hour, how many hours are in a day, and how many days are in a week?

- Discuss how skip counting can be used when counting coins. Be specific.

- Rosa has 5 pennies, 2 nickels, and 4 dimes. Leslie tells her she has she has the same amount as her, however Leslie only has three coins. What coins could Leslie have?

- A nickel is worth more than a dime, since it is larger in size.

- When can you use skip counting in telling time?

- Noon represents nighttime.

- Which period of time is greater: a class that goes from 9:00 a.m. to 10:00 a.m. or an afterschool camp that goes from 3:15 p.m. to 4:15 p.m.? Explain.

Question Quilt

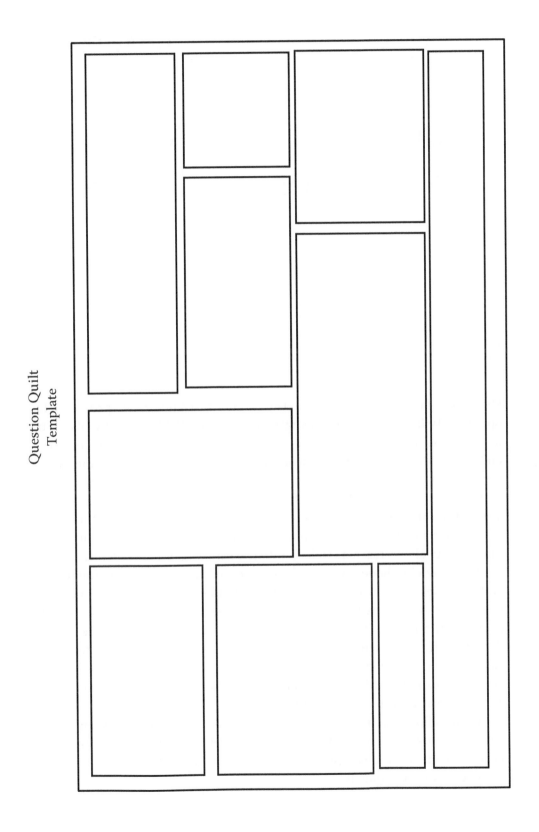

Question Quilt Template

CHAPTER 12

True/False and Always, Sometimes, and Never

Using true/false and always, sometimes, and never questions provides the opportunity for students to investigate a statement. With true and false, students have to determine the validity of the statement. With always, sometimes, and never, students determine whether it is true all the time, some of the time, or never. Typically, always, sometimes, and never have been associated with geometry, but as you can see from the following pages, these are as easily adapted to other topics. They are rich tasks because they engage students in a higher level of reasoning and communication than typical questions might. This activity allows teachers to look into the thinking of the students.

For early childhood students, it is suggested that you work through several examples with the class so they understand this question type and what is expected of them as they explain their answers. Students may not simply explain/write A, S, or N. They must validate and support their choice. If they determine the statement is sometimes, students should be required to offer a case for when the statement does hold (an example) as well as when it does not (a counter example). Many true and false questions can easily be adapted for always, sometimes, and never statements.

In the following, the topical examples start out at a lower level with mostly true/false statements and build to the higher level with always, sometimes, and never statements. You can pick and choose based upon your grade level and then needs of your students.

Sample Statements

Topic 1: Number and Quantity

1. Numbers represent quantities.
2. Six is one more than five.
3. You can only count up.
4. When counting, a two-digit number that ends in a "3" is followed by a two-digit number that ends in a "4."
5. When counting by 5s you only use one column on the hundred chart.
6. On a hundred chart, the even numbers are found in columns.
7. When you subtract you must count backwards.
8. Counting begins with "1."
9. Skip counting is by 1s.
10. You can count up by ones.
11. Sets with different quantities are equal.
12. Any number that ends with three is one more than a number that ends in two.
13. The seven in 37 and 74 represent the same quantity.
14. Six is less than 4.
15. Two tens and twelve ones represent the same quantity as three tens and two ones.
16. Counting backwards can be used to efficiently solve a subtraction situation.
17. In 576, the five can be modeled using five flats from the base ten materials.
18. Two quantities can be both equal and one greater than the other.
19. When you add or subtract 100 from a number, the tens and ones digits will remain the same in the resulting sum/difference.
20. $10 - 10 = 8 - 8$

Topic 2: Algebraic Reasoning

Composing/Decomposing Numbers

1. Composing is subtraction.
2. There are three ways to make 5.
3. You can only determine the sum of two numbers.
4. When adding doubles the resulting sum is even.
5. Doubles plus one results in a sum that is always even.
6. You solve a situation with the total unknown in the same way you solve a situation with an unknown addend.

Operating with Numbers

1. Sums and differences represent the same thing.
2. The order of the addends matters when adding.
3. The order of the quantities matters when subtracting.
4. Adding ten to a number results in a sum that ends in zero.
5. Addition is commutative.
6. Subtraction is commutative.

Models/Equations

1. Equal means the same quantity.
2. The sum is to the right of the equal sign.
3.  can be modeled by $9 = 4 + 5$.
4. Equations have an equal sign.
5. The " $-$ " represents addition.
6. Comparison situations for three quantities are represented using symbols.

Topic 3: Geometric Reasoning and Measurement

Shapes and Geometric Measurement

1. Triangles have four sides.
2. Circles have no sides.
3. Everything can be measured.
4. Flat shapes are 2-D.
5. A house can be below the ground.
6. A shape can be drawn with only two sides.
7. Color is an attribute of a triangle.
8. Trapezoids can be composed using triangles.
9. Rectangles have two sets of sides that are equal lengths.
10. All shapes are flat.
11. The sides of a triangle are the same length.
12. Squares are rectangles.
13. A broken ruler can still be used to measure the length of a book.
14. Rectangles can only be partitioned in half in two ways.
15. Cylinders, cones, and spheres can be stacked together.

True/False and Always, Sometimes, and Never

Time and Money

1. A nickel represents more money than a dime.
2. Time is measured in feet.
3. The colon, ":" is used when telling time.
4. It is 2:30 a.m. in Tennessee and Kentucky, it is daylight.
5. Two quarters have the same value as a half-dollar.
6. Leslie has 13 coins. Tammy has 10 coins. Leslie has more money than Tammy.
7. If three of the coins that Leslie has are pennies, then the amount of money Leslie has will have a "3" in the units place.
8. It is 3:30 a.m., it is daylight.
9. Clocks have two hands.
10. If I have 100 coins and have a total amount of $1.00, I have at least one quarter.

Sample Statements with Answers

Topic 1: Number and Quantity

Counting

1. Numbers represent quantities. **(T)**
2. Six is one more than five. **(T)**
3. You can only count up. **(F)**
4. When counting, a two-digit number that ends in a "3" is followed by a two-digit number that ends in a "4." **(T)**
5. When counting by 5s you only use one column on the hundred chart. **(F)**
6. On a hundred chart, the even numbers are found in columns. **(T/A)**
7. When you subtract you must count backwards. **(S)**
8. Counting begins with "1." **(S)**
9. Skip counting is by 1's. **(N)**
10. You can count up by ones. **(S)**
11. Sets with different quantities are equal. **(F)**
12. Any number that ends with three is one more than a number that ends in two. **(F)**
13. The seven in 37 and 74 represent the same quantity. **(F)**
14. Six is less than 4. **(F)**

15. Two tens and twelve ones represent the same quantity as three tens and two ones. **(T/A)**
16. Counting backwards can be used to efficiently solve a subtraction situation. **(T/S)**
17. In 576, the five can be modeled using five flats from the base ten materials. **(T/A)**
18. Two quantities can be both equal and one greater than the other. **(N)**
19. When you add or subtract 100 from a number, the tens and ones digits will remain the same in the resulting sum/difference. **(A)**
20. 10 − 10 = 8 − 8 **(A)**

Topic 2: Algebraic Reasoning

Composing/Decomposing Numbers

1. Composing is subtraction. **(F)**
2. There are three ways to make 5. **(F)**
3. You can only determine the sum of two numbers. **(T)**
4. When adding doubles the resulting sum is even. **(T/A)**
5. Doubles plus one results in a sum that is always even. **(F/N)**
6. You solve a situation with the total unknown in the same way you solve a situation with an unknown addend. **(S)**

Operating with Numbers

1. Sums and differences represent the same thing. **(F)**
2. The order of the addends matters when adding. **(F)**
3. The order of the quantities matters when subtracting. **(T)**
4. Adding ten to a number results in a sum that ends in zero. **(S)**
5. Addition is commutative. **(A)**
6. Subtraction is commutative. **(N)**

Models/Equations

1. Equal means the same quantity. **(T)**
2. The sum is to the right of the equal sign. **(S)**
3.  can be modeled by 9 = 4 + 5. **(T)**

True/False and Always, Sometimes, and Never

4. Equations have an equal sign. **(T/A)**
5. The " − " represents addition. **(F/N)**
6. Comparison situations for three quantities are represented using symbols. **(S)**

Topic 3: Geometric Reasoning and Measurement

Shapes and Geometric Measurement

1. Triangles have four sides. **(F)**
2. Circles have no sides. **(T)**
3. Everything can be measured. **(F)**
4. Flat shapes are 2-D. **(T)**
5. A house can be below the ground. **(T)**
6. A shape can be drawn with only two sides. **(F/N)**
7. Color is an attribute of a triangle. **(F/N)**
8. Trapezoids can be composed using triangles. **(T/S)**
9. Rectangles have two sets of sides that are equal lengths. **(S [squares are rectangles])**
10. All shapes are flat. **(S)**
11. The sides of a triangle are the same length. **(S)**
12. Squares are rectangles. **(A)**
13. A broken ruler can still be used to measure the length of a book. **(A)**
14. Rectangles can only be partitioned in half in two ways. **(N)**
15. Cylinders, cones, and spheres can be stacked together. **(N)**

Time and Money

1. A nickel represents more money than a dime. **(T)**
2. Time is measured in feet. **(F)**
3. The colon, ":" is used when telling time. **(T)**
4. It is 2:30 a.m. in Tennessee and Kentucky, it is daylight. **(F)**
5. Two quarters have the same value as a half-dollar. **(T/A)**
6. Leslie has 13 coins. Tammy has 10 coins. Leslie has more money than Tammy. **(S)**
7. If three of the coins that Leslie has are pennies, then the amount of money Leslie has will have a "3" in the units place. **(S)**
8. It is 3:30 a.m., it is daylight. **(S)**
9. Clocks have two hands. **(S)**
10. If I have 100 coins and have a total amount of $1.00, I have at least one quarter. **(N)**

PART THREE

Planning and Implementation

CHAPTER 13

Crosswalk

The following crosswalk is included to support instructional planning. Resources can be quickly identified based upon the mathematical topic as well as type of writing and/or the strategy example given. The crosswalk identifies the mathematical topics that are included in the given examples referenced in each of the 11 writing strategies shared in the previous chapter.

Authentic Opportunities for Writing about Math in Early Childhood

Crosswalk of Topics and Writing Strategies

Topics / Writing Strategy	Visual Prompts	Compare/ Contrast	The Answer Is…	Topical Questions	Writing About	Journal Prompts	Poems	Cubing/ Think Dots	RAFTs	Question Quilts	Always, Sometimes, and Never
Number and Quantity	X	X	X	X	X	X	X	X	X	X	X
Algebraic Reasoning	X	X	X	X	X	X	X	X	X	X	X
Geometric Reasoning/ Measurement and Units	X	X	X	X	X	X	X		X	X	X
Data Analysis, Probability and Statistics		X	X	X	X	X					X
Universal						X	X				

140

CHAPTER 14

Bringing It All Together

This last section provides a sample anchor task that demonstrates how several of these writing strategies can be authentically integrated into classroom instruction. A lesson plan and facilitation notes are provided.

Farmer Jones

Overview: The Farmer Jones task was intentionally chosen because it not only models several opportunities for writing but also demonstrates the efficiency of planning for addressing multiple content and process standards. Planning for simultaneous outcomes allows time for students to dive deep into a single task rather than completing multiple tasks over the same timeframe. It allows students to make connections within the content rather than viewing the concepts as discrete and unrelated. This task is so versatile that it can be modified in unlimited ways to meet the needs of both teachers and students.

This task is divided into two parts. The first offers emergent readers and beginning mathematicians opportunities to explore and work with the tangram shapes. Primary students sort shapes and create sorting rules. They also work with tangram puzzles to support building spatial reasoning while they manipulate the shapes. Students further develop their spatial reasoning as they build the tangram square, first with the tangram mat, then later, without the mat. The second part of the task involves measurements and problem-solving with the pieces, their measurements, and beginning work with perimeters. The outline below is a suggestion for how this can be facilitated.

Farmer Jones Tasks:

- Getting to know my tangram pieces (p. 156)
- Tangrams and measurements (p. 161)

Writing Opportunities:

- Topical questions: Here is a beginning set of questions that can be used with students as they are working through the tasks. "We're Stuck/"We're Done" (p. 154)
- Problem-solving process (p. 7, 11, 155)
- Reflection (p. 163)

Content Connections: (This task can be used across grades K-2 so possible topics may include but are not limited to)

- Geometry – Decomposing/Composing,
- Measurement (tangram pieces)
- Problem-solving (standards for mathematical practice)

Materials needed

- Tangrams (Tangrams, in three different colors, can be cut from die cuts and foam, cardstock, construction paper, etc. Alternatively, students can make their own set of tangrams through paper folding and tearing/cutting.)
- Tangram Mat
- Tasks handouts, including suggested questions and reflections
- Venn circles, Hulu hoops, or Venn circle mats for students to use to sort in
- Post-it notes for students to record their sorting rule

PAPER FOLDING A TANGRAM:

- Give the students a square of have them square up a piece of 8.5" x 11" paper.
  - Fold the paper so a shorter side lies on tops of (coincides) with one of the longer sides. Fold back and forth, creasing each time, and tear off the rectangle. You should now have a square piece of paper and a rectangle. Keep the square.
  - Fold the square along one diagonal and crease to make two congruent right triangles. Fold back and forth, creasing each time, and tear apart the right triangles. Set one aside.
  - Take one right triangle and fold in half so you form another set of two congruent right triangles. Fold back and forth, creasing each time, and tear apart the right triangles. Set these aside. They are the first two tangram pieces.
  - Take the second large right triangle and position it so the right triangle is at the top and the hypotenuse is the base. Fold the right angle (the square corner) down to the middle of the opposite side (the hypotenuse). Fold back and forth, creasing each time, and tear apart the small triangle on top from the isosceles trapezoid on the bottom. Set the triangle aside. This is the third tangram piece.
  - Turn the isosceles trapezoid so the longer base is on the bottom. Fold the left side of the trapezoid over on top of the right side so you have folded it in half along a vertical line of symmetry. Unfold and fold the left bottom corner to the middle fold line so the bottom sides lie on top of each other. Crease well and tear apart the small triangle and the remaining small square on the left of the fold line. These are the fourth and fifth pieces of the tangram.
  - Take the remaining trapezoid and turn it so the right angles are on the left and the longer base is on the bottom. Take the upper left corner (at the obtuse angle) and fold it down and to the left corner (the lower right angle) so the bottom sides lie on top of each other. Crease well and tear apart the small triangle on the left and the remaining parallelogram. These are the sixth and seventh pieces of the tangram.

Authentic Opportunities for Writing about Math in Early Childhood

Task 1: Getting to Know My Tangram Pieces

FARMER JONES

Farmer Jones' farm is in the shape of a square. Mrs. Jones loves to play with tangrams. She convinced Farmer Jones to divide his land into pieces just like a tangram puzzle.

Mrs. Jones sorted the pieces for Farmer Jones.

Sort the pieces and be ready to tell how you sorted. Then see if you can sort using a different rule.

FACILITATION NOTES

Making Student Thinking Visible

Students will use tangram pieces and Venn circles to illustrate their sorts. It is important that the pieces be of different colors, so they have the option of at least two sorting rules.

Communication: Students should be able to tell you their sorting rules. They can also write them as they are able.

Bringing It All Together

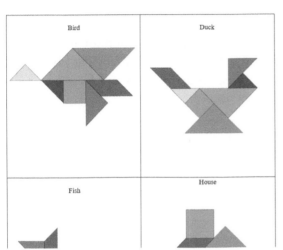

Practice making each of the following shapes with your tangrams.

Bird   Duck
Fish   House

> **FACILITATION NOTES**
>
> **Making Student Thinking Visible**
>
> Students will use tangram pieces and the shape cards as they practice manipulating the various pieces to compose each figure. There are two pages of shapes, with one card being a free choice for the students to create a design of their own making. They will name their design.
>
> **Communication:** As students are composing each figure, they should be able to use appropriate math terms to describe how they made each shape, including if they had to turn/rotate, flip/reflect, etc. Students will describe and name their free choice shape.
>
> **Literacy note:** Any of the following books could be read before or during the study.
>
> - *Grandfather Tang's Story*, by Ann Tompert
> - *Three Pigs, One Wolf*, and *Seven Magic Shapes*, by Grace Maccaron
> - *Warlord's Puzzle*, by Virginia Pilegard

Authentic Opportunities for Writing about Math in Early Childhood

FACILITATION NOTES:

Making Student Thinking Visible

These task cards that can be used as students continue to explore tangram manipulations. Note: When you put together two triangles of the same size, you can make a square. This property is not found in all triangles. It works here since the tangram triangles are right triangles.

Communication: As students manipulate the tangram pieces, monitor their discussions for proper vocabulary. As able, students can record their work in their Mathematician's Notebook as journal entries, using drawings and /or sentences to show their thinking.

Trace and name each of your tangram pieces.	Make a square using only one tangram piece.	Make a square using two tangram pieces.
What shapes can be put together to make a triangle?	What shapes can be put together to make a trapezoid?	What shapes can be put together to make a rectangle that is not a square?

Bringing It All Together

FARMER JONES

Farmer Jones' farm is in the shape of a square. Mrs. Jones loves to play with tangrams. She convinced Farmer Jones to divide his land into pieces just like a tangram puzzle.

Can you make the square for Farmer Jones with the tangram pieces?

FACILITATION NOTES

Making Student Thinking Visible

It is suggested that some students would benefit from first making the square using the tangram mat. Then, they can try it without the mat.

Communication

Students should be able to identify the specific challenges they faced in completing this task. For example, they kept turning a piece to make it fit, when it needed to be flipped over/reflected.

Getting to Know My Tangram Pieces: We're Stuck/We're Done Questions

1. How many pieces are there in the tangram?
2. What can you tell me about these shapes?
3. What was your first sorting rule?
4. What was your second sorting rule?
5. Mrs. Jones sorted the tangram pieces so there were three different sets. How do you think she sorted the pieces?
6. What will happen if you rotate the triangle?
7. What will happen if you flip the parallelogram?
8. How must we move this shape in order to make it fit?
9. How will your work mat aid you in working with this task?

Authentic Opportunities for Writing about Math in Early Childhood

Tangram Master

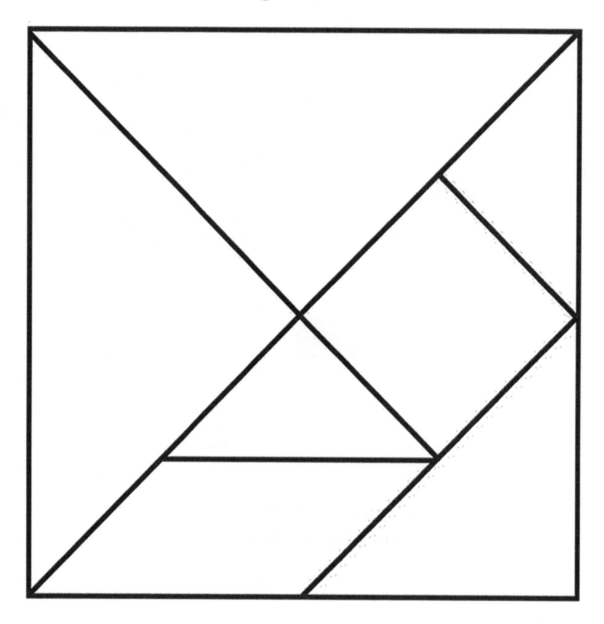

Bringing It All Together

Task 2: Tangrams and Measurement Task

FARMER JONES

Now that Farmer Jones' farm has divided his land into pieces just like a tangram puzzle, he needs to fence off different areas.

Using your tangram pieces, measure each of the sides for each piece. Sketch the pieces and record their measurements in the chart below.

Tangram Piece	Sketch with Measurement (units)
Small triangle	
Square	
Medium triangle	
Parallelogram	
Large triangle	

Order the pieces from smallest perimeter to greatest perimeter.
Facilitation notes:

Making Student Thinking Visible

Students can measure using non-standard units or traditional standard/metric units. You can determine what best meets your students' needs.

Communication: If students, in the same class, used different units, they should be able to compare and contrast the relationships between the various side lengths of the shapes. Students could also order based on what they perceive the overall "size" to be, even though they have not yet worked with area. Be sure students can articulate these relationships as they are foundational for the next part of the task.

Authentic Opportunities for Writing about Math in Early Childhood

Tangrams and Measurements: We're Stuck/We're Done Questions

1. How long is each side of the completed tangram square?
2. Measure each side of all the tangram pieces.
3. What is the relationship between each of the pieces of the tangram?
4. How do you know?

Farmer Jones has decided how he is going to use each piece of the tangram. He asked Mrs. Jones to buy fencing for each of the areas. Help Mrs. Jones figure out how much fencing she needs to order.

- The square/ house has a side length of 100 feet.
- The barn/pond small triangle has a long side that is 140 feet long.
- The cows/large triangle has short sides with a length of 200 feet and the long side being 283 feet.

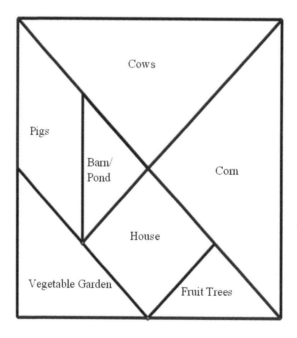

Bringing It All Together

FACILITATION NOTES:

Making Student Thinking Visible/Communication

Have students complete the Problem-Solving Process Graphic Organizer. See Chapter 1.

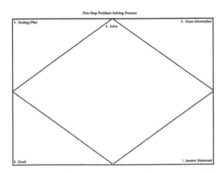

When students get to part 2, monitor them carefully. Be sure they can identify what is given as well as what they still need to know.

In Part 3, as students identify the strategy they will use, they may need to think ahead to part 4.

In Part 4, students should be able to articulate the relationships they identified in the previous part of this task to be able to determine the remaining side lengths.

Note: Lengths can be varied to allow for the current quantitative limits of your students.

The task can be extended to include a variety of fencing. For example, Mrs. Jones wants white picket fence around the house, the fruit trees, and the vegetable garden. Mr. Jones wants wire fence around the cow field and the corn field and wood slat fence around the pig and barn areas.

For students who are ready, the cost of the fence could be determined.

Getting to Know My Tangram Pieces

FARMER JONES

Farmer Jones' farm is in the shape of a square.
Mrs. Jones loves to play with tangrams.
She convinced Farmer Jones to divide his land into pieces just like a tangram puzzle.
Mrs. Jones sorted the pieces for Farmer Jones.
Sort the pieces and be ready to tell how you sorted.
Then see if you can sort using a different rule.

Practice making each of the following shapes with your tangrams.

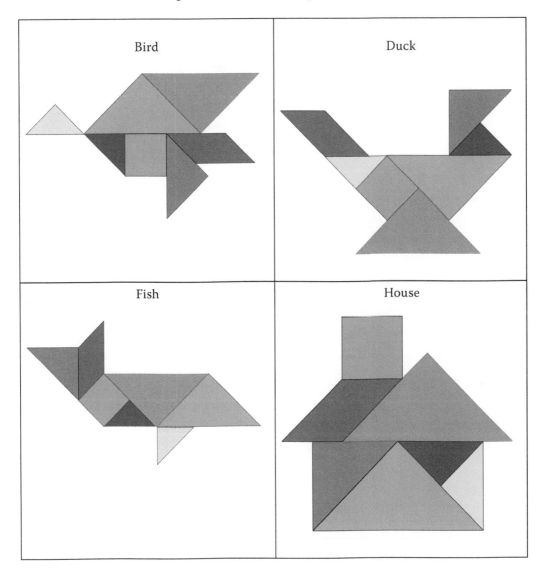

Authentic Opportunities for Writing about Math in Early Childhood

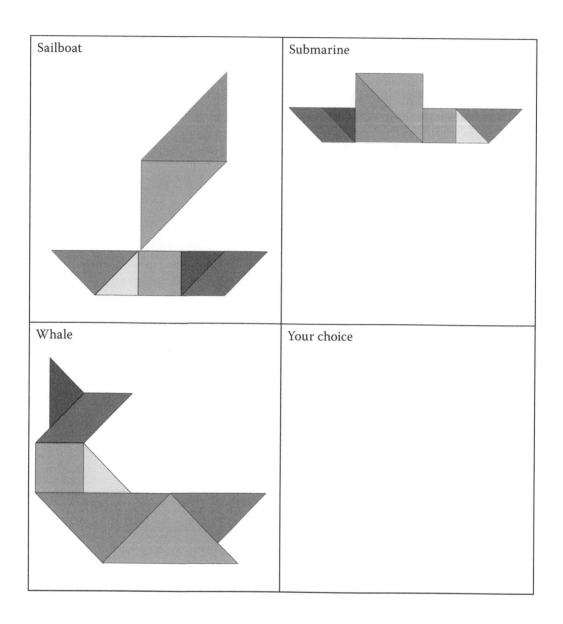

Bringing It All Together

Trace and name each of your tangram pieces.	Make a square using only one tangram piece.	Make a square using two tangram pieces.
What shapes can be put together to make a triangle?	What shapes can be put together to make a trapezoid?	What shapes can be put together to make a rectangle that is not a square?
What shapes can be put together to make a parallelogram?	Make a square using three tangram pieces.	Make a square using four tangram pieces.
Make a square using five tangram pieces.	Make a square using six tangram pieces.	Make a shape of your choice.

Authentic Opportunities for Writing about Math in Early Childhood

FARMER JONES

Farmer Jones' farm is in the shape of a square.

Mrs. Jones loves to play with tangrams.

She convinced Farmer Jones to divide his land into pieces just like a tangram puzzle.

How can you make the square for Farmer Jones with the tangram pieces?

Bringing It All Together

Tangrams and Measurement Task

FARMER JONES

Now that Farmer Jones' farm has divided his land into pieces just like a tangram puzzle, he needs to fence off different areas.

Using your tangram pieces, measure each of the sides for each piece. Sketch the pieces and record their measurements in the chart below.

Tangram Piece	Sketch with Measurement (units)
Small triangle	
Square	
Medium triangle	
Parallelogram	
Large triangle	

Order the pieces from smallest perimeter to greatest perimeter.

Farmer Jones has decided how he is going to use each piece of the tangram. He asked Mrs. Jones to go buy fencing for each of the following. Help Mrs. Jones figure out how much fencing she needs to order.

- The square/ house has a side length of 100 feet.
- The barn/pond small triangle has a long side that is 140 feet long.
- The cows/large triangle has short sides with a length of 200 feet and the long side being 283 feet.

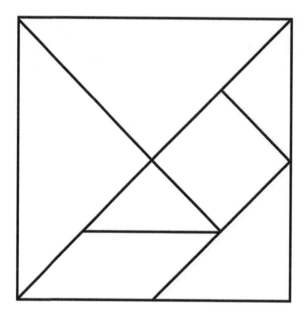

Bringing It All Together

QUAD Reflection

Q: Question:
What questions do you still have about the task?

U: Understanding
What do you now understand after working with this task?

A: Activate:
How did this task activate you as a learning resource for a peer or a peer for you?

D: Discourse:
What discourse did working this task prompt?

QUAD Reflection

Q: Question:
What questions do you still have about the task?

U: Understanding
What do you now understand after working with this task?

A: Activate:
How did this task activate you as a learning resource for a peer or a peer for you?

D: Discourse:
What discourse did working this task prompt?

Afterword

We have always had educators ask us if they could get a list of the questions we used during the day's training, a list of the writing prompts we used, an example of a DI strategy we mentioned, etc. Our questions came from our interactions with the participants; the writing prompts, while mostly intentional, also might have been inspired by a comment or question from a participant, and, yes, we did have examples of the various strategies that we often mention. At the same time, our editor, Lauren, to whom we owe so much thanks and gratitude, asked us if we would ever consider doing a book of consumables for educators to use. Hence, the seed was sown that finally grew into what became this book series.

Everyone acknowledges that communicating in mathematics is essential. Communication was one of the original five process standards of The National Council of Teachers of Mathematics. Over the years, we have collected, found, and created many classroom resources that provide authentic opportunities for students to communicate mathematically. So, the next step was culling through the many resources we had used and developed over the years. We sorted, resorted, looked at, accepted some, rejected others, and even created new ones as needed. We wanted to offer this series in four books to meet the individual needs of the various grade bands. At the same time, we wanted to provide examples and prompts that would cover the breadth of the grade

Afterword

band's mathematical topics and provide materials to support deepening the understanding of the topics.

Then to the writing and pulling together of the resources, which for us as mathematicians, the latter was much less challenging. This is ironic since this book focuses on writing and communicating in mathematics! What emerged is what you have on these pages. We hope that this resource becomes a go-to to meet your everyday classroom needs for providing opportunities for your students to engage in communicating about mathematics and communicating mathematically. Take what we have provided, expand on it, and make it your own. As you reimagine, retool, and even create your versions, we ask that you reach out and share. We would love to hear from you. You can find us at https://tljconsultinggroup.com/about-us/tammy-jones/ and https://leslietexasconsulting.com/about-leslie/.

Bibliography

Growney, J. *Intersections—poetry with mathematics.* https://poetrywithmathematics.blogspot.com

Jones, T. L. and Texas, L. A. (2017). *Strategic journeys for building logical reasoning: Activities across the content areas.* New York: Routledge, Taylor & Francis Group.

Smith, M. S. and Stein, M. K. (2022). *5 Practices for orchestrating productive mathematics discussions.* Reston, VA: National Council of Teachers of Mathematics, Inc.

Southern Regional Education Board. (2018). *Making math matter: High-quality assignments that help students solve problems and own their learning.* SREB. https://www.sreb.org/sites/main/files/file-attachments/18v04_math_matters_report_final.pdf?1521473373

Texas, L. A. and Jones, T. L. (2013). *Strategies for common core mathematics: Implementing the standards for mathematical practice.* New York: Routledge, Taylor & Francis Group.